Reconfigurable RF Power Amplifiers on Silicon for Wireless Handsets

ANALOG CIRCUITS AND SIGNAL PROCESSING

Series Editors: Mohammed Ismail
Mohamad Sawan

For other titles published in this series, go to
http://www.springer.com/series/7381

Laurent Leyssenne · Eric Kerhervé · Yann Deval

Reconfigurable RF Power Amplifiers on Silicon for Wireless Handsets

Springer

Laurent Leyssenne
IMS Laboratory
33405 Talence
France
laurent.leyssenne@yahoo.fr

Eric Kerhervé
IMS Laboratory
33405 Talence
France
eric.kerherve@ims-bordeaux.fr

Yann Deval
IMS Laboratory
33405 Talence
France
yann.deval@ims-bordeaux.fr

ISBN 978-94-007-0424-4 e-ISBN 978-94-007-0425-1
DOI 10.1007/978-94-007-0425-1
Springer Dordrecht Heidelberg London New York

Printed on acid-free paper

Springer is part of Springer Science+Business Media (www.springer.com)

Preface

The design of power amplifiers (*PA*) for wireless applications has been a topic of great concern among the scientific community in electronics since the early 1990s. Power amplifiers dominantly determine the power efficiency and battery lifetime of modern mobile terminals. *PA* linearity is a key feature that limits the maximum allowed data rate of a radio link. The deployment of always more complex radio networks with a continuously growing throughput sharpens the specifications of power amplifiers. The market of wireless handsets (cellular phones, ...) pushes manufacturers towards multi-standard capabilities (3G, Bluetooth, *WIFI*) and a growing level of integration. The *Bill-of-Material* and the consumed die area that are affordable are constantly reduced for cost purpose. The trade-offs that are inherent to *PA* design are therefore extremely stringent and no technical/technological solution can unanimously be regarded as a definitive contribution. To this date, the market of handset-dedicated power amplifiers is widely dominated by *III/V* technologies. However, throughout the following pages, we will try to highlight the benefits of *PA* integration on silicon. The architectures that will be proposed hereunder take advantage of silicon capabilities and strength, among others their relatively low cost and their ability to combine high power devices with low-power analog/digital control circuitry.

Chapter 1 will first present the respective features of 2nd and 3rd generation cellular applications (*GSM, DCS, EDGE, WCDMA...*) and data transmission standards (*WIFI, WIMAX, LTE*). An overview of the most commonly employed *RF* power amplifier topologies will also be provided, with their advantages and drawbacks. Finally, the 0.25 μm *BICMOS* ST Microelectronics technology will be described and compared with *III/V* processes in the prospect of *PA* development. The fundamental features of power devices will be detailed and the most appropriate technological choice prior to the *PA* design itself will be discussed.

In Chapters 2 and 3, several novel *PA* topologies will be proposed and discussed in terms of efficiency, linearity and complexity by means of mixed system/transistor-level analyses. Chapter 2 will investigate three novel switched-mode power amplifier topologies. The first non-constant-gain principle is based on the power stage bypass/extinction and applied to a silicon *HBT* demonstrator. A silicon *HBT* demonstrator that was developed in the frame of *RNRT ASTURIES* project will be presented The other two topologies were proposed in the frame of the European Medea+ *UPPERMOST* project and deal with the dynamic control of a fragmented reconfigurable multi-cell power stage. First, open-loop power stage control will be considered. In this topology, power detection is carried out by a specifically dedicated digital ΔΣ *Built-In Current Sensor* whose behavior will be detailed. Second, a closed-loop power stage control system will be addressed and compared with the open-loop power stage control.

Lastly, Chapter 3 will describe a *PA* Module that was developed in the frame of the European *FP6 MOBILIS* project. Furthermore, some mathematical developments and vector illustration will explain the theory of a continuously and dynamically power adaptive system that aims to combine linearity and efficiency even at low power levels. Moreover, frequency-dependent memory effects will be introduced in order to provide some further insight in the complex non-linear phenomenon's that may alter the behavior of power amplifiers operating at wide channel bandwidths. An illustration of memory effects will be provided in the experimental section of this chapter. A *PA* module demonstrator based on integrated silicon and passive network dice will be presented, both in a stand-alone mode and in association with the other blocks of the *MOBILIS* transceiver. The final discussion will validate the proposed efficiency/linearity improvement principle under some conditions of channel bandwidth.

Talence, France

Laurent Leyssenne
Eric Kerhervé
Yann Deval

Acknowledgments

This work was realized at IMS Laboratory, Bordeaux, France, in collaboration with ST Microelectronics. We would like to express our gratitude to Mr. Didier Belot (RF Design Manager at ST Microelectronics, Crolles, France), Mr. Daniel Saias (*CEO* at Asygn, Montbonnot, France, and formerly manager of the RF systems and architecture team at ST Microelectronics, Crolles, France) and Mr. Hilal Ezzeddine (RF Design Manager at ST Microelectronics, Tours, France) who respectively provided the necessary support for the development, processing of silicon and passive *IPD* dice, as well as the access to ST Microelectronics test and characterization facilities in Crolles and Tours. The authors would also like to thank Mr. Patrice Gamand (RF Innovation Center General Manager at NXP Semiconductors, Caen, France) and Mr. Andreas Kaiser (CNRS Research Director at IEMN, Lille, France) for revising the technical validity of this monograph as well as Daniel Rees Lewis (ENSEIRB, Bordeaux, France) for his support on its linguistic correctness. Other thanks to Mrs. Magali de Matos (Engineer at IMS Laboratory, Bordeaux, France) for her help in *on-boar*d and *on-wafer* characterizations of stand-alone integrated circuits and *PCB* demonstrators.

Contents

Abbreviations

A

ACLR	Adjacent Channel Leakage Ratio
ADC	Analogue to Digital Converter
AGC	Automated Gain control
AM/AM	(Input) Amplitude to (output) Amplitude conversion
AM/PM	(Input) Amplitude to (output) Phase conversion

B

BAW	Bulk Acoustic Wave (resonator)
BER	Bit Error Rate
BICS	Built-In Current Sensor
BOM	Bill Of Material
BPSK	Binary Phase Shift Keying
BV	Breakdown Voltage
BV_{CE}	Collector/Emitter breakdown voltage
BV_{CB}	Collector/Base breakdown voltage
BV_{DS}	Drain/Source breakdown voltage

C

CCDF	Complementary Cumulative Distribution Function
CDMA	Code Division Multiple Access
CHE	Channel Hot-Electron injection
CMFB	Common-Mode Feed-Back
COB	Chip on Board
CW	Continuous-Wave

D

DAC	Digital to Analogue Converter
DACH	Drain Avalanche Hot Carrier injection

DCS	Digital Communication System
DPCCH	Dedicated Physical Control Channel
DPDCH	Dedicated Physical Data Channel
DSP	Digital Signal Processing
DUT	Device Under Test

E

EDGE	Enhanced Data rates for *GSM* Evolution
EER	Envelope Elimination and Restoration
ET	Envelope Tracking
EVM	Error Vector Magnitude

F

FBI	Feed-Back Information
FCC	Federal Communications Commission (USA)
FDD	Frequency Division Duplex

G

GBW	Gain×Bandwidth product
GMSK	Gaussian Minimum-Shift Keying
GPRS	General Packet Radio Service
GSM	Global System for Mobile communications

H

HEMT	High Electron Mobility Transistor
HBT	Hetero-junction Bipolar Transistor
HICUM	HIgh Current transistor Model
HPSK	Hybrid Phase Shift Keying
HS-DPDCH	High-Speed Dedicated Physical Data Channel
HSUPA	High-Speed Uplink Packet Access

I

IF	Intermediate Frequency
IMD_3	3rd order Inter-Modulation Distortion
IPD	Integrated Passive Device
ISI	Inter-Symbol Interference

L

LDD	Lateral Lightly Doping
LDMOS	Laterally Doped MOSFET
LINC	LInear amplification with Non-linear Components
LNA	Low Noise Amplifier
LO	Local Oscillator
LOCOS	LOCal Oxidation of Silicon

N

NF	Noise Figure
nodeB	Base Station
NTF	Noise Transfer Function

O

OCP_1	1 dB Output Compression Point
OED	Oxidation Enhanced Diffusion
OFDM	Orthogonal Frequency Division Multiplexing
OIP_3	3rd order Output-referred Intercept Point
OSR	Over-Sampling Ratio

P

PA	Power Amplifier
PAE	Power Added Efficiency
PAPR	Peak-to-Average Ratio
PCB	Printed Circuit Board
PLL	Phase-Lock Loop
PSK	Phase-Shift Keying
PWM	Pulse Width Modulation

Q

QAM	Quadrature Amplitude Modulation
QOS	Quality of Service
QPSK	Quadrature Phase Shift Keying

R

RNRT	Réseau National de Recherche en Télécommunication (French National Research Network in Telecommunication)
RRC	Root Raised Cosine

RSSI	Received Signal Strength Indicator
RX	Reception path

S

SIP	System in Package
SNR	Signal-to-Noise Ratio
SNDR	Signal-to-Noise/distortion Ratio
SOI	Substrate on Isolator
STF	Signal Transfer Function
STI	Shallow Trench Isolation

T

TDD	Time Division Duplex
TDMA	Time Division Multiple Access
TFCI	Transport Format Channel Information
TPC	Transport Power Control
TQFP	Thin Quad Flat Pack
TX	Transmission path

U

UE	User Equipment

V

VCO	Voltage-Controlled Oscillator
VGA	Variable-Gain Amplifier
VSWR	Voltage Standing-Wave Ratio

W

WCDMA	Wide-band Code Division Multiple Access
WIMAX	Worldwide Interoperability for Microwave Access
WLAN	Wireless Local Area Network

Chapter 1
Mobile Phone Transmitters for Wireless Standards: Systems, Architectures and Technologies

Abstract This chapter aims to paint a broad picture of the technological limitations that power amplifiers for wireless handsets must cope with. In order to maximize spectral efficiency and immunity to fading/interferers/noise, modern RF standards generally use techniques such as Spread spectrum or Orthogonal-Frequency-Duplex methods. Their drawback is increased dynamic envelope variations. To preserve linearity, power amplifiers are sometimes forced to operate in a backed-off regime whereby their efficiency is reduced. The most commonly used efficiency-enhancement architectures are reviewed here while their respective advantages/drawbacks are assessed from the standpoint of complexity, bill-of-material, die area and compliance with integration on silicon. At last, the capabilities of the power devices that are available in a 0.25 μm *BICMOS SiGe* technology (ST Microelectronics) are discussed comparatively to *III/V* processes in terms of power gain, linearity, and robustness to thermal runaway and/or output load mismatch.

Keywords Power amplifiers (*PA*) · 3G/4G RF standards · Spread spectrum · *WCDMA* · *OFDM* · Efficiency enhancement · Envelope Elimination and Restoration (*EER*) · Envelope Tracking · Silicon technology · Laterally Doped MOS (*LDMOS*) · Heterojunction Bipolar Transistor (*HBT*) · Integrated Passive Device technology

1.1 RF Cellular/Data Transmission Standards and Related Handset Uplink Architectures

1.1.1 Introduction

The goal of this introduction is not to give an exhaustive description of cellular and *Wireless Local Area Network* (*WLAN*) standards but give only the key elements that help the understanding of the most critical issues and trade-offs in uplink transmitters and power amplifiers (*PA*). The battery lifetime of a handset/terminal is greatly determined by the *PA* efficiency, and, in order to increase it, *TX* and/or *PA* architectures must be adapted according to the addressed *RF* standards/modulation schemes. This topic is further explored in the following section.

L. Leyssenne et al., *Reconfigurable RF Power Amplifiers on Silicon for Wireless Handsets*, Analog Circuits and Signal Processing, DOI 10.1007/978-94-007-0425-1_1,

1.1.2 Second Generation Radiofrequency Standards and Their Implication on Uplink Architecture

Second generation *GSM/GPRS* standards are based on Gaussian Shift Keying modulation (*GMSK*). They are only based on phase modulation as depicted in Fig. 1.1. Such a constant envelope modulation scheme makes the design of an efficient transmitter relatively easy. A *GSM/GPRS* handset *TX* is traditionally implemented by means of a phase-lock-loop (*PLL*) controlled by a $\Delta\Sigma$ modulator that provides digital phase information (1: +90° shift; 0: –90° shift). The power amplifier efficiency is maximized by the use of a switching class topology (class E or F). *PA* design has two key constraints. *RMS* phase error should not exceed 5°. As *GSM* operates in time duplex mode (*TDD*), burst ramp-up/down must fulfill a temporal mask (i.e. limited On/Off settling times). In order to enhance the throughput of *GSM/GPRS*, a compliant extension known as *EDGE* was introduced on top of it. Although this standard implies low impact on network and base-band processing levels, it features a non-constant-envelope modulation scheme based on 8-*Phase Shift Keying* (Fig. 1.1).

Consequently, the *GSM/EDGE* compliant transmitter architecture had to be updated and transformed into a (open-loop) *Polar Modulated* or a *Polar (closed-) Loop* structures, respectively depicted in Figs. 1.2 and 1.3. In parallel to additional blocks being committed to control the output envelope magnitude, phase is processed in the same way it used to be for a *GSM/GPRS* transceiver. This will be detailed in the section dedicated to the *Envelope Elimination and Restoration* (*EER*, see further, Section 1.2.3.1), an efficient power amplifier topology that is directly inherited from polar modulation. In a polar loop, the output *RF* signal is first down-converted. Then the output phase is probed by a limiter, compared to the initial reference phase and the resulting phase error is fed-back to the *PLL*. In the same way, a power detector probes the output magnitude. The magnitude error resulting

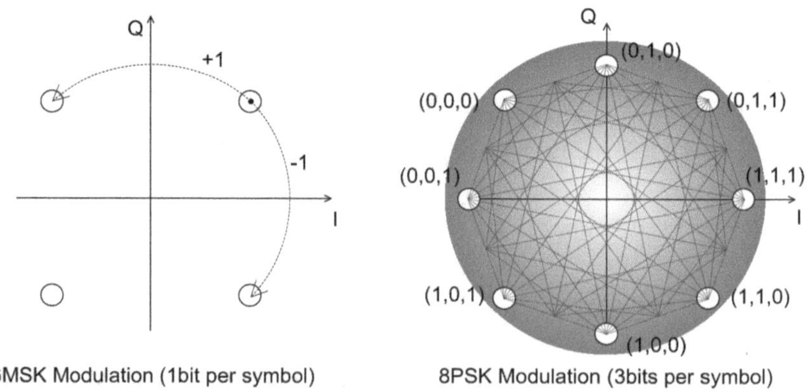

Fig. 1.1 Illustration of *GMSK* and 8*PSK* modulation schemes

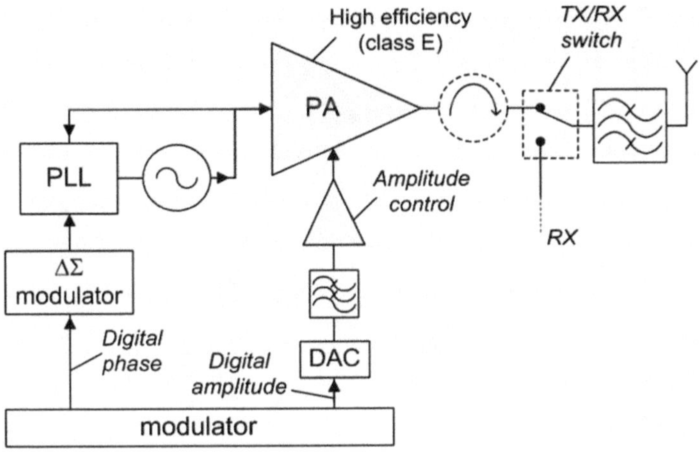

Fig. 1.2 Polar modulated architecture

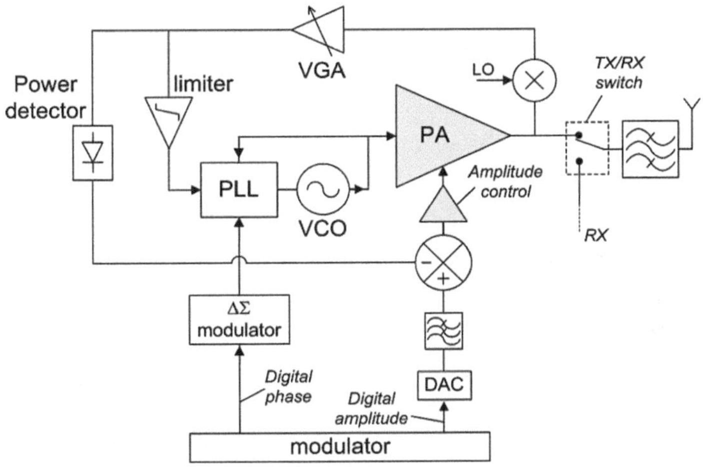

Fig. 1.3 Polar loop architecture

from the comparison with the reference magnitude (provided by the modulator), is fed-back to the *Amplitude Control* amplifier. Though more complex than polar modulation, polar loop is a much more robust topology to antenna mismatch (i.e. to *Voltage Standing Wave Ratio*) and makes the use of an isolator obsolete.

Linearity specifications are expressed either in terms of time burst mask (*GSM/EDGE*), or of phase error (*GSM*) or in terms of *Error Vector Magnitude* (*EDGE*). Spectral requirements must also be fulfilled according to stringent spectral masks as depicted in Fig. 1.4.

The properties and specifications of *GSM/EDGE* are summarized in Table 1.1.

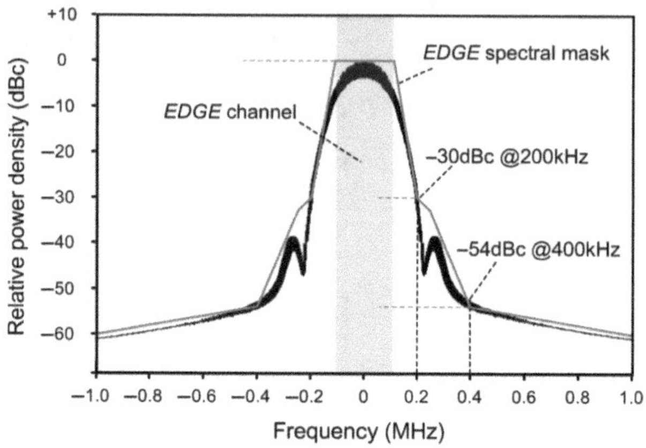

Fig. 1.4 *EDGE* spectral mask specification

Table 1.1 *GSM/EDGE* properties and specifications

		GSM	*EDGE*
(Quad) bands (MHz)		[880–915] [925–960] [1710–1785] [1805–1880]	
Modulation scheme		*GMSK*	8*PSK*
Duplexing method		*TDD*	
Channel spacing		200 kHz	
Bit rate (kbps)		271	813
Max. output power (dBm)		33	26
Spectral mask (dBc)	@200 kHz	−30	
	@400 kHz	−60	−54
	@1800 kHz	−63 (Low Bands), −65 (High bands)	
Phase error (°)		5 (*RMS*), 20 (peak)	–
EVM (%)		–	9 (*RMS*), 30 (peak)

1.1.3 Cellular Third Generation CDMA-Based Standards

1.1.3.1 Currently Used 3G Standards

Third generation standards that are currently deployed worldwide are divided into two families: *CDMA*2000 (North America) and *W-CDMA* (Europe) (3GPP 1999, Holma and Toskala 2007). They are both based on *Code-Division Multiple Access* (*CDMA*). Contrary to *GSM*/*EDGE* networks that allocate communication timed scheduled slots to users (*Time Division Multiple Access*), *CDMA* access method allows several users to simultaneously transmit/receive several data channels in the same frequency band (Fig. 1.5).

In order to overcome the *Near-Far* problem, 3rd generation networks are designed so as to manage soft handover (between one terminal and several base stations) and accurate power control (± 1 dB). Indeed, if such power control was not carried out, base stations and user equipments might operate at full-power when not

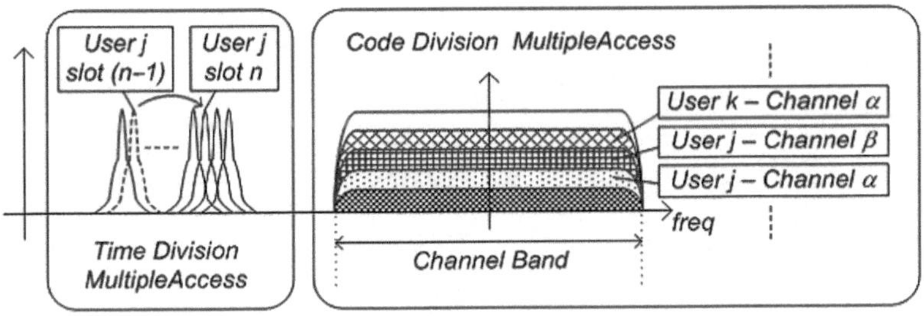

Fig. 1.5 *TDMA* vs. *CDMA*

needed which would lead to power inefficiency and/or to network overload. Power control must be able to accommodate for output power variations due to circuit performance deviation (for example process. . .), environmental conditions (antenna *VSWR*, temperature amongst others) and multi-path fading. Furthermore, the range over which power control must be applied is much broader for third generation standards than for *GSM/EDGE* (30/50 dB). This means that 3G transmitters may operate at very low power levels for which high noise/efficiency issues are observed. A common way of addressing those problems is to switch the *PA* off under stringent conditions on power gain discontinuity (see Section 1.1.3.5).

Power control is implemented by means of two methods:

- *Closed-loop power control*
  Within this method, user terminals (*UE*) and base stations (*nodeB*) are involved in a closed-loop system that allows the uplink/downlink power levels to be maintained at the appropriate value according to the *UE/nodeB* distance, the traffic conditions and the cell capacity (Fig. 1.6).
  The time response of this closed-loop must be low enough to compensate for fast multi-path delays within a cell, both in pedestrian and vehicular scenarios. Practically, closed-loop time response should be in the order of 1 μs.
- *Open-loop power control*
  The expression "Open-loop" is a little ambiguous and must be understood from the point-of view of the *UE/nodeB* communication. One practical way to

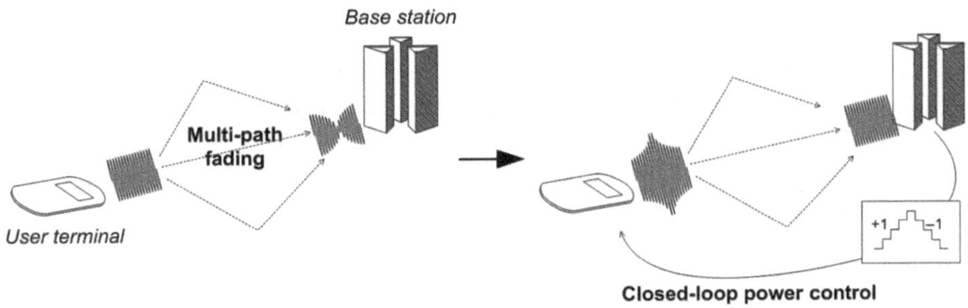

Fig. 1.6 Illustration of closed-loop power control

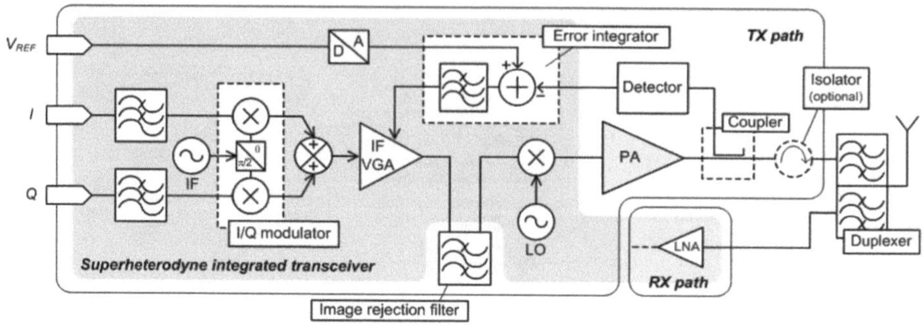

Fig. 1.7 Super-heterodyne transmitter architecture, including automated gain control loop

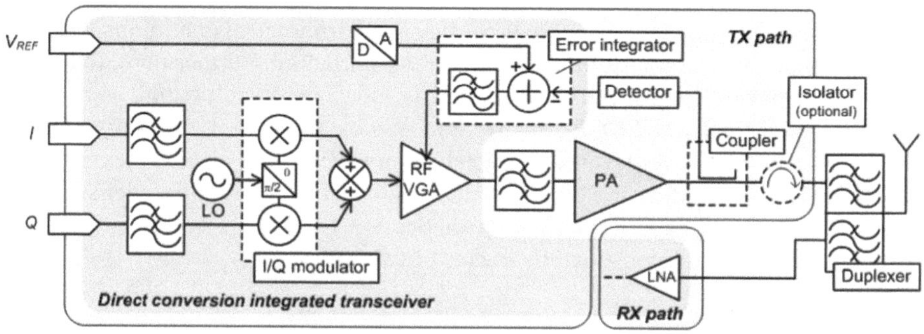

Fig. 1.8 Direct conversion transmitter architecture, including automated gain control loop

implement this kind of control is to insert an *Automated Gain Control* loop (*AGC*) in the transmitter. Typical uplink architecture for 3G standards including an *AGC* loop are depicted in Figs. 1.7 and 1.8. Third generation standards use *QAM* modulation schemes which makes the use of polar loop impractical. Therefore, super-heterodyne or direct conversion *I/Q* architectures are most commonly used.

For purposes of reduced power consumption, and *PCB* area, direct conversion is often utilized. The insertion of selective band-pass filter prior to *PA* is then necessary in order to loosen the noise/linearity requirements on the modulator and the receive path.

The *AGC* feedback calculates the average error envelope by means of a power detector and a very narrow-band low-pass filter (or an integrator) and controls a *Variable Gain Amplifier* (*VGA*). A logarithmic detector in association with a linear-in-dB *VGA* is sometimes used to shorten the loop time response and enhance the detection dynamic range.

Another feature of third generation standards is their non-constant envelope modulations, and such *I/Q* transmitter architectures do not alleviate the traditional linearity/efficiency trade-off. Therefore, *PA* design techniques must be employed to preserve battery life-time without increasing *Bit-Error Rate* (*BER*).

1.1.3.2 Spread Spectrum Technique: Principle, Advantages and Limitations

Channel capacity, i.e. the maximum achievable data rate in a very noisy environment is given by Shannon & Hartley's law:

$$C = \frac{BW}{\ln(2)} \cdot \ln(1 + SNR) \underset{SNR<<1}{\approx} 1.44 \cdot BW \cdot SNR \tag{1.1}$$

where BW is the channel bandwidth and SNR is the signal-to- noise ratio.

At a given signal-to-noise ratio, channel capacity can be enhanced by increasing the bandwidth over which data is distributed. In addition, the spread spectrum technique that is used by third generation standards consists in increasing the channel capacity by convoluting the data spectrum with that of wide-band pseudo-noise sequences. These spreading sequences, (based on Walsh or Hamadard codes in practice) build an orthogonal base that is associated to a correlation scalar product. Correlation consists in computing the following integral expression (1.2):

$$C(s_1, s_2, t_0) = \frac{1}{T_S} \cdot \int_{t_0}^{t_0+T_S} s_1(t) \cdot s_2^*(t)\,dt \tag{1.2}$$

where s_1 and s_2 are signals to correlate, T_S is the symbol period.

The correlation of a spread signal with the corresponding spreading code that was previously employed allows the determination of the original data (providing synchronization acquisition is properly achieved), or otherwise returns to zero. In this way, spectrum spreading allows the discrimination of several users, which guaranties confidentiality and provides immunity against blockers and noise. In the uplink, spread spectrum also allows the separation of several information channels that are associated to different spreading codes. The longer the spreading sequence, the higher the processing gain and the immunity to noise.

The corollary benefits of spread spectrum are that it provides some frequency diversity and immunity to multi-path fading as long as this latter's effect is narrow-band enough comparatively to channel bandwidth. Indeed, the various delayed signals resulting from a unique source can be individually separated from the others at the receiver by means of specifically-delayed *correlation fingers*. They are further on coherently combined to rebuild the original non-faded signal. Such functionality requiring *multi-correlation fingers* is known as a Rake receiver and is typical of spread spectrum systems. The larger the network cell size, the higher the rake complexity. That is a limitation of the spread-spectrum technique.

1.1.3.3 Description of European 3G: *WCDMA* and Its Uplink High Data-Rate Extension *HSUPA*

The generation of a *Wideband CDMA* (*WCDMA*) signal follows the principle described hereunder. Once processed by several digital operations (coding, redundancy bit insertion and interleaving), the uplink bit stream is split into several channels that carry distinct information:

- The *Dedicated Physical Control Channel* (*DPCCH*) provides the base-station receiver with the necessary information for a good radio link. This information is the *Pilot* that carries the channel and signal-to-interference ratio estimates (determined by the handset receiver), the *Transport Format Channel Information (TFCI)* that contains the data rate information, the *Transport Power Control* (*TPC*) that informs the base-station whether it must increase/decrease its downlink power, and the *Feedback Information* (*FBI*) that controls base station *beamforming*. As this channel plays an essential role, it requires robust despreading and maximum processing gain. Spreading factor is set to 256.
- One or several *Dedicated Physical Data Channel* (*DPDCH*) contain the data to be transmitted. Data rate may be updated on every frame, from 15 up to 960 kbps. Consequently the associated spreading factor may vary from 4 to 256.

In order to be discriminated by the base-station receiver, *DPDCH* and *DPCCH* channels are spread by two distinct Walsh codes, featuring identical 3.84 Mcps chip rate but potentially different lengths. This chip rate value makes *WCDMA* more robust under fading than *CDMA*2000 (1.25 Mcps) and alleviates Rake receiver complexity. *WCDMA* frame is structured as depicted in Fig. 1.9.

The uplink power is updated in every slot, i.e. every 666 μs. Indeed, when all or part of the stages are bypassed and switched-off, for the purpose of efficiency/noise figure improvement, severe discontinuities might be observed in the power gain magnitude and phase. Even if the open loop power control partially compensates for the magnitude variations, closed-loop power control is necessary to ensure a proper power control at ± 1 dB, as specified by *WCDMA*.

Therefore, the theoretical limit rate $F_{RECONFIG}$ at which *hard reconfiguration* of the power amplifier may be applied is therefore given by the slot duration as follows:

$$F_{RECONFIG} = \frac{1}{\Delta\tau_{slot}} = \frac{1}{666 \cdot e - 6} = 1.5\,\text{kHz} \tag{1.3}$$

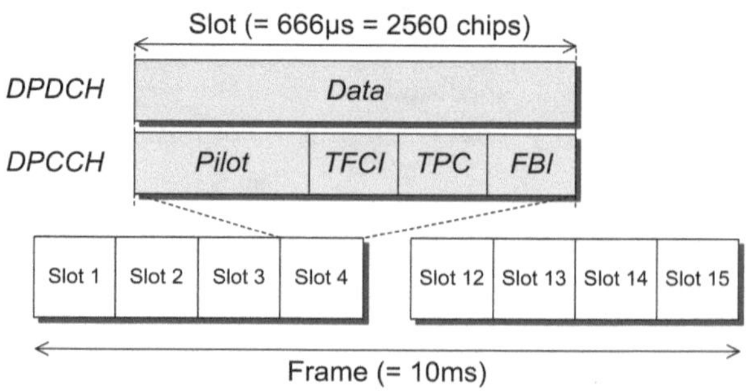

Fig. 1.9 *WCDMA* frame structure

Since open/closed power control only handles signal magnitude and not phase, a more restrictive empirical law was included in *WCDMA* release 5 specifications as a function of the phase hopping $\Delta\theta_{STEP}$ due to *PA* reconfiguration:

The channels *DPDCH* and *DPCCH* are respectively mapped to *In-phase* and *Quadrature* signals (resp. *I* and *Q*) of a dual-channel *QPSK* constellation. Depending on traffic conditions (i.e. data rate or the number of data channels), the weight of the control channel (sqrt(G_C)) can be adaptively reduced so as to improve spectral efficiency. The constellation shape is therefore effected and may adaptively move from *QPSK* ($G_C = 1$) nearly to *BPSK* ($G_C \sim 0$) as depicted in Fig. 1.10a. This also adaptively increases the peak to average power ratio (*PAPR*). To overcome this issue, *WCDMA* is based on a *HPSK* modulation scheme (Hybrid Phase shift keying). *HPSK* modulation is quite similar to *QPSK* in that the constellation is built on an alternately-rotated dual *QPSK* constellations. Both *QPSK* constellations are rotated one from the other by an angle that is adaptively changing according to *DPCCH* weight, as depicted in Fig. 1.10b. When $G_C = 1$, both constellations overlap. For every symbol period, the bit stream is alternately mapped to one or the other *QPSK* constellation. Thus, over a symbol duration, only 90° transitions are possible. Symbol transitions are the only potential occurrences of 180° phase-steps for which the signal locus may return to approximately zero. This means that *HPSK* modulation significantly reduces the *Peak to Average Ratio*, comparatively with dual-channel *QPSK* (downlink modulation scheme). The *Complementary Cumulative Distribution Function* value (*CCDF)* is the probability that output power exceeds the average value by a given *PAPR* headroom in dB. $PAPR_{HPSK}$ is approximately 3.1 dB @$CCDF = 0.1\%$, whereas $PAPR_{QPSK}$ is 6.8 dB @$CCDF = 0.1\%$.

Constellation alternate rotation is carried out by means of a complex-valued scrambling operation. To this end, spreading codes and pseudo-noise scrambling codes are employed. The goal of scrambling in the uplink is to discriminate terminal

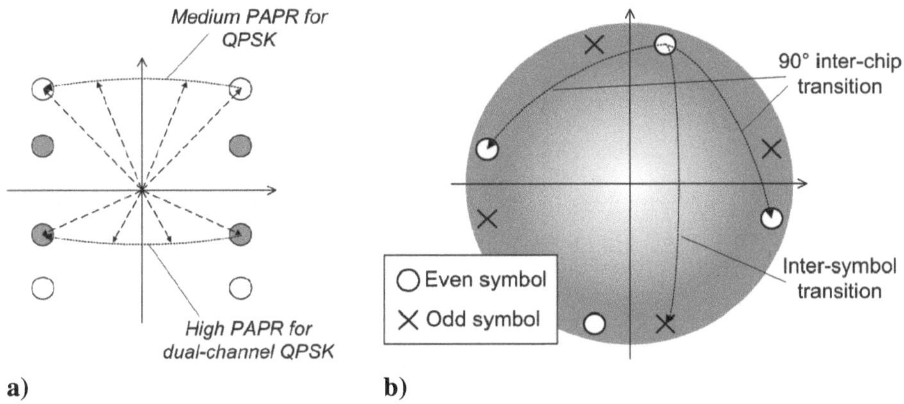

Fig. 1.10 (**a**) Constellation with a dual-channel *QPSK* modulation scheme (*PAPR* increase), (**b**) Constellation with a *HPSK* modulation scheme (*PAPR* reduction)

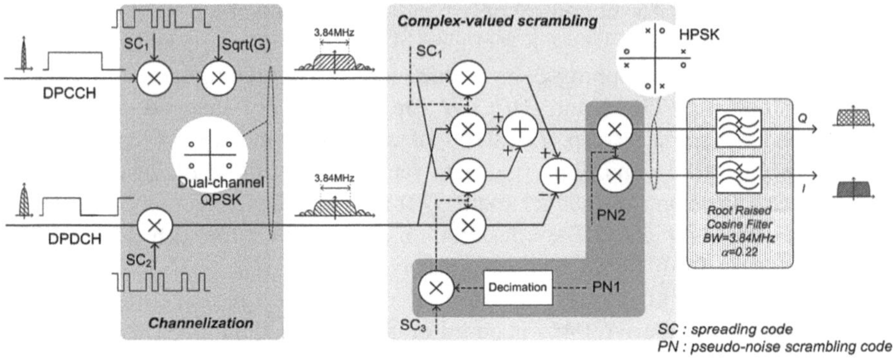

Fig. 1.11 *WCDMA* base-band processing

users with identical spreading codes in the same network cell. The overall base-band processing that is required by *HPSK* modulation scheme is summarized in Fig. 1.11.

Low-pass shaping filters are implemented according to *Root-Raised Cosine* topology (*RRC*) with a 0.22 roll-off factor. Their goal is to enhance spectral efficiency/capacity. The drawback of this highly selective filtering is that it contributes to an additional increase of *PAPR* and *Inter-Symbol Interferences* (*ISI*).

1.1.3.4 *WCDMA* Linearity Requirements

The 60 MHz wide *TX WCDMA* band is split into eleven channels featuring a 5 MHz-spacing. The immunity of one channel to his neighbors is of great importance and impacts the required transmitter linearity. This constraint is expressed by means of a constellation-based specification, the *error vector magnitude* (*EVM*):

$$EVM = \sqrt{\frac{\sum_{j=1}^{N} \left(I_{out,j} - I_{ref,j}\right)^2 + \left(Q_{out,j} - Q_{ref,j}\right)^2}{N \cdot P_{AVG}}} \tag{1.4}$$

where N is the number of chips in the frame, P_{AVG} the average output power and by a spectrum-based specification, the *adjacent channel leakage ratio* (*ACLR*):

$$ACLR = 10 \cdot \log \left(\frac{power_{adjacent\_channel}}{power_{main\_channel}} \right) \tag{1.5}$$

The power levels that are involved in the computation of *ACLR* are integrated over 3.84 MHz-wide bands. A 5 MHz spacing between the main and the adjacent channels is assumed. An illustration of *ACLR* determination is given in Fig. 1.12

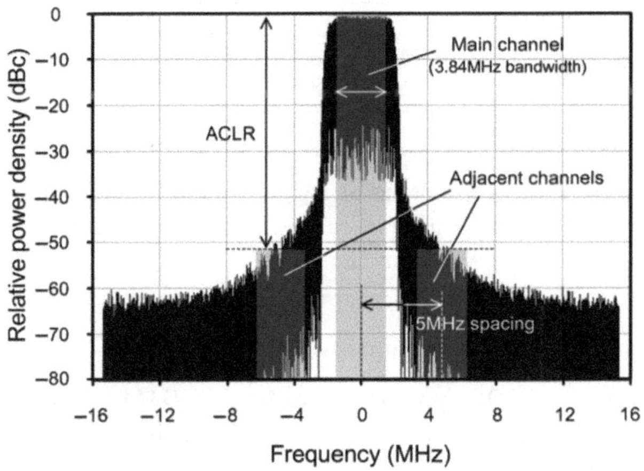

Fig. 1.12 *ACLR* illustration on a *WCDMA* spectrum

1.1.3.5 *WCDMA* Dynamic Range

WCDMA dynamic range that is induced by the near/far problem and multi-path fading is specified as high as 74 dB (in contrast to 30 dB for *GSM*). The uplink power range is therefore [−50 dBm ; 24 dBm]. At $P_{OUT} = -50$ dBm, the efficiency and noise issues are critical. Most *PA* efficiency enhancement techniques are generally developed and optimized at high power levels but do not address the efficiency issue at low or medium power levels. In this prospect, in order to relax the modulator noise specification and reduce the transmitter consumption, the functionality of *PA* bypass is sometimes necessary as depicted in Fig. 1.13 (*PA* power gain here is assumed to be 30 dB). When low power levels are required at the antenna level, the power amplifier is switched-off and bypassed. The constraints on modulator dynamic range are then relaxed by the same amount as the *PA* gain. Such technique implies that the Automatic Gain Control quickly compensates for sudden *TX* gain

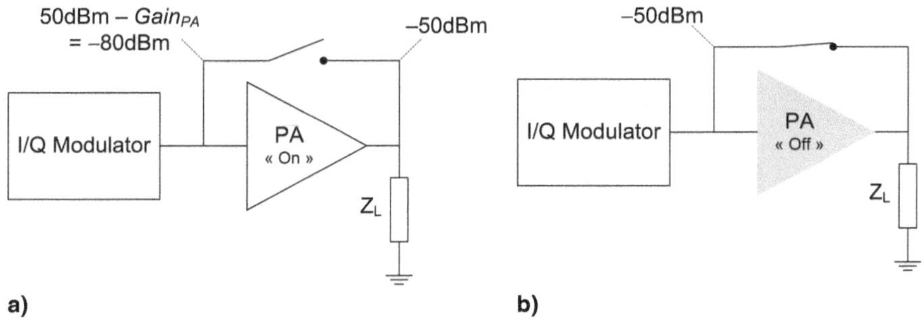

Fig. 1.13 Illustration of *PA* bypass interest in *WCDMA* minimal power scenario: (**a**) noise sensitive configuration with active *PA*; (**b**) noise-relaxed configuration with bypassed *PA*

Table 1.2 Maximum allowed hard reconfiguration rate as a function of phase discontinuity

$\Delta\theta_{STEP} < 30°$	$F_{RECONFIG} = 1.5$ kHz
$30° < \Delta\theta_{STEP} < 60°$	$F_{RECONFIG} = 0.3$ kHz
$\Delta\theta_{STEP} > 60°$	Not allowed

variations. Moreover, the phase steps in *TX* gain should be minimized for purpose of compliance with the specification in Table 1.2. A *SiGe PA* demonstrator based on this principle is described in Chapter 2.

1.1.3.6 *WCDMA* Noise Requirements

One of the drawbacks of *FDD* applications such as *WCDMA* lies in the noise leakage from transmit to receive paths. In order to relax *RX* sensitivity (*LNA* noise figure for example), it is necessary to keep this noise leakage below a specified level that is determined by the duplexer *TX/RX* isolation. The noise budget at the front-end level can be summarized in Fig. 1.14. The values that are indicated in Fig. 1.14 are typical of most developed transmitters and were also used in the frame of *MOBILIS* project that will described in Chapter 3 (MOBILIS 2006).

where NF_{BB} is the noise figure of the *RX* base-band block,
NF_{RF} is the noise figure of the *RX RF* front-end (*LNA*, band-pass filter, mixer...),
NF_{PA} is the noise figure of the TX power amplifier module,
N_{TX/RX_LNAin} is the maximum acceptable noise level at the *LNA* input due to *TX*,
N_{TX/RX_PAout} is the maximum acceptable noise level at the *PA* output,

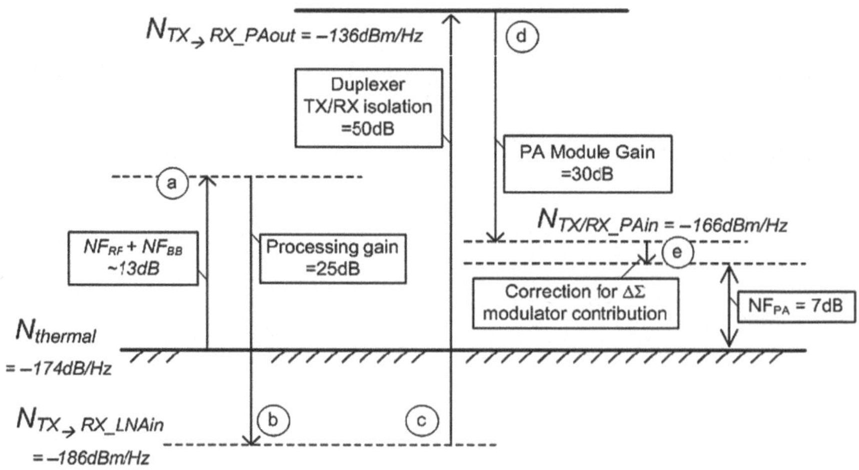

Fig. 1.14 Methodology of noise budget determination in a *FDD* spread spectrum application

N_{TX/RX_PAin} is the maximum acceptable noise level at the *PA* input,
$N_{thermal}$ is the −174 dBm/Hz thermal noise floor (i.e. −108 dBm over a 3.84 MHz *WCDMA* channel).

In the case of a 12.2 kbps *WCDMA* application (voice-only data rate), the processing gain provided by spread spectrum is:

$$G_{process\_dB} = 10.\log\left(\frac{3.84e^{+6}}{12.2e^{+3}}\right) \approx 25\text{dB}$$

Within *MOBILIS* project, the targeted RX noise figure and duplexer *TX*/*RX* isolation are respectively 13 and 60 dB. The minimum noise level that is allowed by the *RX* path if the *TX* path was noiseless is (−174 + 13 − 25) = −186 dBm/Hz. This value represents the maximum limit that the noise leakage from *TX* to *RX* should not exceed. The maximum noise power density at the *PA* output is therefore −126 dBm/Hz which implies a maximum −61 dBm noise power over the 3.84 MHz channel bandwidth.

1.1.3.7 Uplink *WCDMA* Extension: High Speed Uplink Packet Access

In order to improve the throughput of uplink *WCDMA* up to 5.8 Mbps, without increasing chip rate, a channel *High Speed-Dedicated Physical Data Channel (HS-DPDCH)* was introduced on top of the *DPDCH*. The consequence of this is a more dense/complex constellation, and an increase of *peak-to-average ratio* up to approximately 7 dB @*CCDF* = 0.1%. *HSUPA* is therefore an evolution that has negative effect on *TX* front-ends since it degrades efficiency. It necessitates either *PA* redesign or additional *PA* back-off, i.e. it reduces the maximum linear power that can be transmitted by an already designed *PA*.

1.1.3.8 *WCDMA* Specification Summary

The requirements of uplink *WCDMA* are summarized in Table 1.3

Table 1.3 Summary of *WCDMA* properties

	Uplink *WCDMA*
Multiplexing method	*CDMA*
Duplexing mode	*FDD*
Modulation scheme	*HPSK*
Channel bandwidths (MHz)	5
Bit rate (Mbps)	1
Spectrum bands (MHz)	1920–1980
Max output power (dBm)	24 (class 3)
ACLR (dBc)	−33
RMS EVM (%)	17.5

1.1.4 Data Transmission Wireless Standards

1.1.4.1 Generalities About *OFDM* Based Standards

Modern data transmission wireless standards are all based on the *OFDM* division method. This acronym stands for *Orthogonal Frequency Division Multiplexing*. This technique is employed by many wireless applications such as:

- IEEE 802.11a,g,n *WLAN* and *HiperLAN*
- IEEE 802.16d,e (fixed and mobile *WIMAX*), *WIBRO*
- DVB-T,H (fixed and mobile Digital Video Broadcast)
- 3GPP *LTE* (long term evolution)

The *OFDM* division technique allows spectral efficiency to be enhanced (up to ~4 bps/Hz for *WIMAX*) and therefore quality of service compared with *TDMA* (0.5 bps/Hz for *GSM*) or *CDMA* (2 bps/Hz for *HSDPA*) is higher. Another interesting *OFDM* feature is its low sensitivity to blockers or multi-path fading. This is all the more important as many of the previously mentioned standards are specified to operate in unregulated *ISM* bandwidths, especially at 2.45 and 5.8 GHz, where numerous interferer sources may be encountered (microwave ovens, Bluetooth, amongst others). Such techniques are based on the calculation of complex Fourier transforms (*FFT*) and Inverse Fourier transforms (*IFFT*) which requires huge *digital signal processing* capabilities. Therefore, the deployment of *OFDM* technology was only made possible in the past few years thanks to the development of nanoscale *CMOS* processes.

1.1.4.2 *OFDM* Principle

The *OFDM* principle consists in mapping a data stream to several *orthogonal* sub-carriers. Orthogonality must be understood in that the correlation product of two distinct sub-channels over a symbol period is nil (see (1.2) for details) and auto-correlation of a single sub-channel gives 1. The general synoptic of an *OFDM* transmitter is depicted in Fig. 1.15. Binary data are first coded (scrambling,

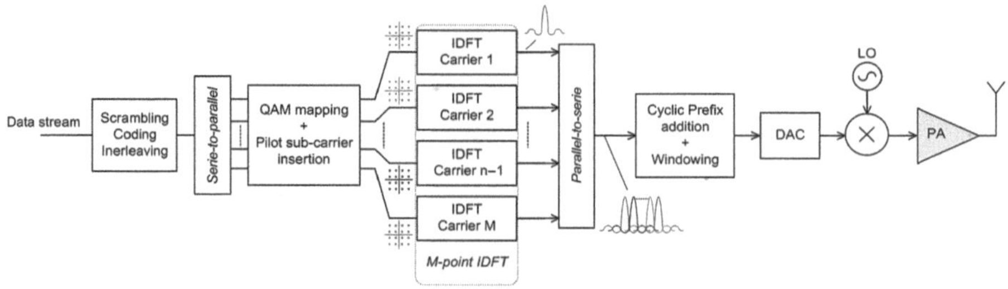

Fig. 1.15 *OFDM* transmitter architecture

Reed-Solomon, Viterbi, interleaving, amongst others) for safety and confidentiality purpose. Then the data is parallelized into "slow" symbols that are mapped to a constellation whose complexity can be adaptively configured according to traffic conditions (from *BPSK* to 64 *QAM*). Pilot symbols are inserted in order to facilitate synchronization at the beginning of the radio link as well as equalization. Each parallel signal modulates a sub-channel in the digital domain via an *IFFT* computation block. All digital complex sub-channels are then summed, converted into an analogue (*I,Q*) signal, and up-converted to the *RF* carrier frequency.

OFDM does not use spectral shaping but presents low sensitivity to inter-symbol interference by inserting an interval guard. Both *OFDM* and spread spectrum techniques can be combined, e.g. in the case of *WIFI* that uses *frequency-hopping spectrum spreading.* The computation of *Error Vector Magnitude* is more complex for *OFDM* than for spread spectrum since several sub-carriers must be taken into account. *EVM* is calculated with the following expression:

$$EVM = \frac{1}{N_f} \cdot \sum_{i=1}^{N_f} \sqrt{\frac{\sum_{j=1}^{L_P} \left[\sum_{k=1}^{M_{SC}} \left(I_{out}(i,j,k) - I_{ref}(i,j,k)\right)^2 + \left(Q_{out}(i,j,k) - Q_{ref}(i,j,k)\right)^2 \right]}{L_P \cdot M_{SC} \cdot P_{AVG}}} \tag{1.6}$$

Where N_f is the number of frames in the transmitted information, L_P is the number of symbols in a frame and M_{SC} is the number of sub-channels.

1.1.4.3 *OFDM* Robustness to Multi-Path Fading

Multi-path fading generally has a narrow-bandwidth destructive effect and is of concern only for a limited number of sub-channels/symbols. Transmission immunity to fading can therefore be improved by adaptively selecting the most efficient sub-channels. Contrary to *CDMA* applications, *WLAN* and *WIMAX* draw advantage from frequency diversity and feature an *Orthogonal Frequency Division Multiplex Access* method. *OFDMA* allows dynamically allocating groups of sub-carriers to various users according to the radio link quality. Sub-carriers for which severe power reduction are encountered due to fading can be adaptively left apart. In order to enhance resilience to fading, *OFDM* is often employed in combination with *Multiple-In Multiple-Out* technology (*MIMO*). Such a technique consists in duplicating transmitter and receiver front-ends. *MIMO* architecture draws advantage from spatial diversity, i.e. it reduces the occurrence probability of strong destructive interference that may result in sharp notches over the emitted spectrum. *MIMO* can also be used to implement *Beam Forming* by adaptively controlling the phase shifts of the various *TX/RX* paths. To this date, *MIMO* is generally not supported by mobile terminals/handsets due to its high area consumption and complexity.

1.1.4.4 *OFDMA* vs. *SC-FDMA*

One of the fundamental drawbacks of *OFDMA-based* applications compared to *spread-spectrum-based* signals is that they present very high *PAPR* (up to 11 dB at $CCDF = 10^{-3}$). This makes the design of a linear and efficient power amplifier very challenging. This is all the more critical as the overall occupied bandwidth is very broad (10–20 MHz). This drawback has motivated extensive research in the domain of efficient access methods (Myung 2006).

Hence, *Single-Carrier Frequency Division Multiplex Access* (*SC-FDMA*) was chosen as the standardized uplink access method for *Long Term Evolution* (*LTE*) application. *SC-FDMA* is similar to *OFDMA* in that it divides the transmission bandwidth into smaller sub-carriers and uses guard intervals. Identically to *OFDMA*, *SC-FDMA* equalization in the reception path is carried out in the frequency domain. The synoptic of a *SC-FDMA* transmitter is given by Fig. 1.16. Minor changes are observed by comparison with *OFDMA* except the insertion of an additional digital *DFT* processing prior to the sub-carrier mapping. Such access method can therefore be considered as a *DFT-spread OFDMA*. *SC-FDMA* present several advantages over *OFDMA*. One such advantage is it increases robustness to spectral nulls and sensitivity to carrier frequency offset. On top of that, it reduces *PAPR* down to 6 dB (resp. 8 dB) at $CCDF = 0.1\%$ for a QPSK (resp.16-*QAM*) modulation scheme. This alleviates the linearity/efficiency trade-off and is a strong argument in favor of *LTE* deployment.

1.1.4.5 Summary of *WLAN*, *WIMAX* Main Characteristics

The characteristics of uplink Mobile *WIMAX* and *WLAN* are summarized in Tables 1.4 and 1.5. *WIMAX* and *WLAN* present very stringent linearity requirements in terms of spectral mask and *EVM*.

The spectral masks are very dependent on local policies and on standard releases. Indeed, 802.11a/b/g/n, 802.16d/e (d and e respectively stand for fixed and mobile) are not treated identically, as it can be noticed on Fig. 1.17 (North America *FCC* specifications).

EVM specifications for *WIMAX* and *WLAN* are summarized in Table 1.6a and 1.6b.

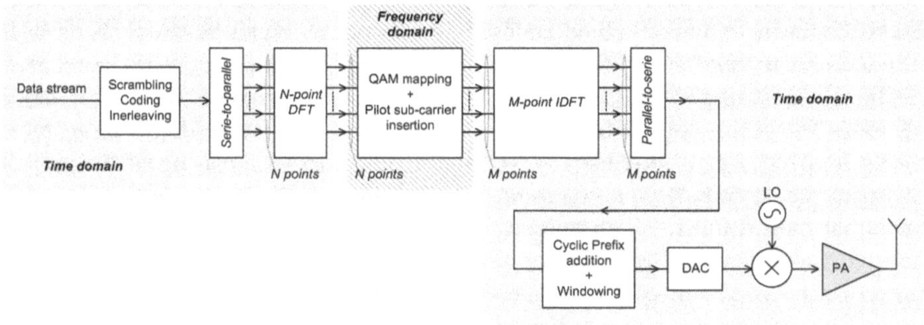

Fig. 1.16 *SC-FDMA* transmitter architecture

Table 1.4 *WIMAX* characteristics

	Mobile *WIMAX*
Standard	IEEE 802.16e
Multiplexing method	*OFDMA*
Duplexing mode	*TDD*
Modulation scheme	*QPSK*, 16-*QAM*, 64-*QAM*
Signal bandwidths (MHz)	5, 7, 10
FFT size	Scalable
Sub-carrier spacing (kHz)	10.94
Spectrum bands (GHz)	2.3, 2.5, 3.5

Table 1.5 *WLAN* characteristics

	WLAN
Standard	IEEE 802.11x
Multiplexing method	*OFDM*
Duplexing mode	*TDD*, *FDD*
Modulation scheme	*BPSK*, *QPSK*, 16-*QAM*, 64-*QAM*
Signal bandwidths (MHz)	20
Spectrum bands (GHz)	2.45 and 5.5

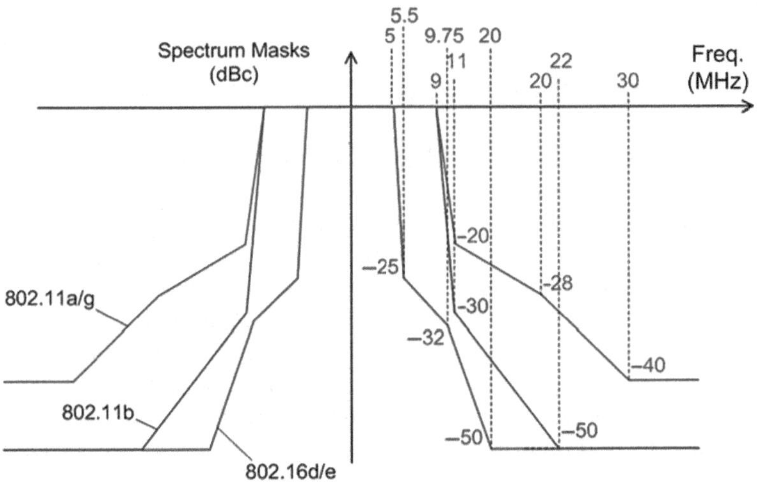

Fig. 1.17 Spectral masks of the main *OFDM* applications

Table 1.6a *EVM* specifications for *WIMAX*

Modulation scheme	*EVM* for 802.16e (%)
QPSK (3/4)	11.9
16-*QAM* (3/4)	5.6
64-*QAM* (3/4)	2.8

Table 1.6b *EVM* specifications for *WLAN*

Modulation scheme	*EVM* for 802.11x (%)
36 Mbps	11.2
54 Mbps	5.6

1.1.5 Power Back-Off Determination

In order to reduce the time-to-market required by a new *RF* platform, engineers often have to cope with the design of *TX* transceivers while standard properties are not fully regulated and stabilized. As long as software tools do not support the latest release of modulated source libraries, the accurate response of a power amplifier with the actual modulated signal cannot be simulated. To tackle the issue of the envelope swing and make sure that a power amplifier will not operate in a non-linear regime, the targeted *CW* performance is assessed by adding a power headroom (generally known as *back-off*) to the specified average power. To the past few years, the computation of this power back-off was based on the *PAPR*. This rule of thumb was found to be relevant for low data-rate spread-spectrum signals but little suitable for *HSUPA* and *OFDM* signals. Indeed, experimental analyses on various *TX* demonstrators using *OFDM* technique have demonstrated that their power capability was sometimes overestimated. To overcome this limitation, a more reliable back-off estimation, the *Cubic Metric* (3GPP 2006), was therefore necessary and is now widely regarded as a safer theoretical tool.

1.2 Power Amplifier Topologies for User Equipment

1.2.1 Introduction on Power Amplifiers Typical Issues

As previously mentioned, average power in modern wireless standards features broad dynamic range. On top of that, modulation-specific envelope swing accentuates the instantaneous power range over which a power amplifier must be able to efficiently operate according to stringent spectral and *EVM* requirements and under antenna mismatch conditions. This briefly summarizes the numerous challenges that *PA* design nowadays implies. The next paragraphs draw on a review of the *PA* topologies that are traditionally used in literature to overcome the efficiency/linearity trade-off.

1.2.2 Base Stations Dedicated Efficiency Enhancement PA Architectures

1.2.2.1 Outphasing Power Amplifiers

Architectures based on outphasing power amplifiers (also referred to as *LINC* architectures) are sometimes employed to implement high efficiency *PA* for base stations. The fundamental principle consists in transforming a magnitude and phase-modulated signal *S*(*t*) into two phase-only-modulated signals $S_+(t)$ and $S_-(t)$ according to (1.7):

$$\begin{aligned} S_+(t) &= S(t) + E(t) = A_{\max} \cdot \exp(\varphi(t) + \varphi(t)) \\ S_-(t) &= S(t) - E(t) = A_{\max} \cdot \exp(\varphi(t) - \varphi(t)) \end{aligned} \tag{1.7}$$

where $S(t) = A(t) \cdot \exp(\varphi(t))$,
$E(t) = jA(t) \cdot \sqrt{\frac{A_{\max}^2}{|A(t)|^2} - 1} \cdot \exp(\varphi(t))$, $\varphi(t) = \cos^{-1}\left(\frac{A(t)}{A_{\max}}\right)$, and $A_{\max}$ is the maximum voltage swing $A_{\max} = \max(A(t))$.

In other words, the amplitude information is converted to phase information. The geometric illustration of this transformation is depicted in Fig. 1.18. As the *out-phased* signals $S_+(t)$ and $S_-(t)$ do not carry any magnitude information, efficient amplification becomes easily possible by means of switching class topologies (class D, E or F). Indeed, those *PA* topologies are the most appropriate candidate to maximize *power added efficiency* (*PAE*) or *drain/collector efficiency* (η), important efficiency figures of merit defined as follows:

$$PAE = \frac{P_{OUT} - P_{IN}}{P_{DC}}\ ;\ \eta = \frac{P_{OUT}}{P_{DC}} \tag{1.8}$$

where P_{OUT} and P_{IN} are respectively the output and input power of the *RF* signal and P_{DC} is the average consumed power.

At the output level, additive combination of $S_+(t)$ and $S_-(t)$ allows the quadrature component *E*(*t*) to be canceled and the original magnitude modulation is revealed. The overall architecture is represented in Fig. 1.19.

Several on-chip topologies were proposed for the separation component prior to power amplifiers (Hammes 2004, Shi 2000). Their design requires special care since tiny phase imbalances result in significant distortion. The output combiner is

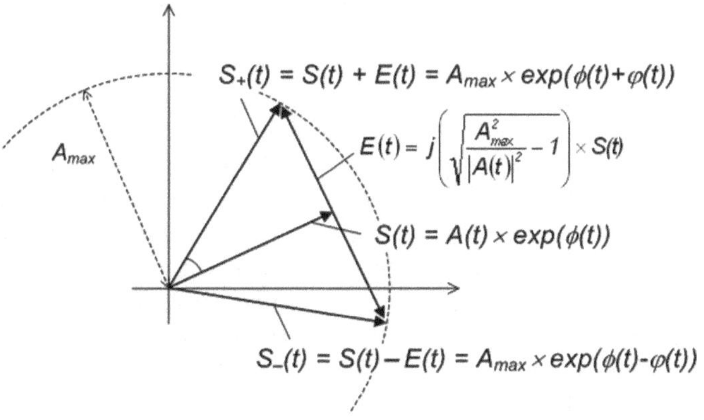

Fig. 1.18 Illustration of outphased separation

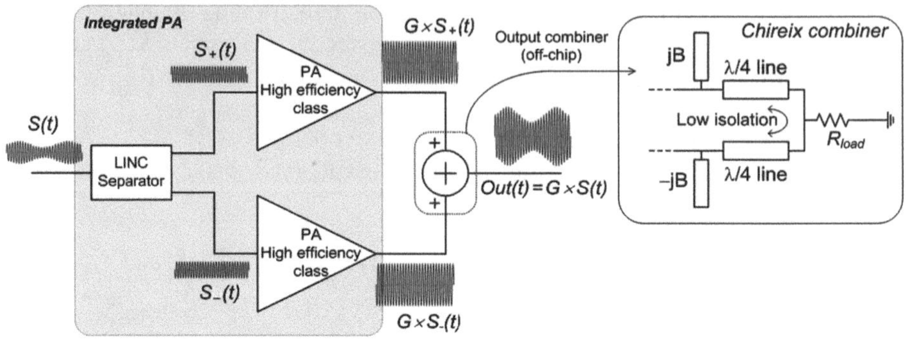

Fig. 1.19 Outphasing amplifier architecture

also a critical element in the whole architecture. The traditional combiner topology is the *Chireix* combiner (Chireix 1935). It presents low loss and high efficiency but contributes to the increase of distortion. This is due to the low isolation from one *PA* to the other that results in time varying impedances and unexpected phase shifts at the combiner ports. Phase predistortion is therefore desirable (Birafane and Kouki 2005). Alternative hybrid combiner with power recycling was also proposed by (Zhang et al. 2002) that circumvents linearity problems at the expense of low efficiency loss.

Combiners can hardly be integrated on silicon and are left off-chip. This makes *LINC* architecture area inefficient at 2 GHz. *LINC* is generally considered for the design of base-station *PA* and is little compliant with the design of cellular handsets.

1.2.2.2 Doherty Architecture

The traditional Doherty principle is fundamentally based on active load-pull technique (Doherly 1936, Raab 1987). A linear amplifier and a weakly biased "auxiliary" amplifier are coherently combined via a quarter-wave line whose characteristic impedance is twice the load, as depicted on Fig. 1.20. A Doherty design method is given in (Sirois et al. 2005).

At low power level, the auxiliary amplifier is nearly switched-off and contributes little to the output signal. Its output impedance is high enough to be considered as an *open* circuit. The linear (or carrier) amplifier is responsible for the class AB amplification. The power added efficiency quickly increases with power. Whereas the linear amplifier starts saturating above a power threshold, the auxiliary counterpart gradually gets biased, thus increasing the power capability of this architecture. Efficiency is then maintained at a high level with no excessive distortion. Indeed, the non-linearity that arises from the auxiliary amplifier activation can be overcome by proper compensation with that of the linear carrier amplifier. This is especially the case with *field effect transistors* devices and some explanations will be given about this in the technology-oriented section (see Section 1.3.2.2).

One of the traditional Doherty limitations lies in the relatively low power gain and low power range over which it operates (6 dB). Some works have proven the

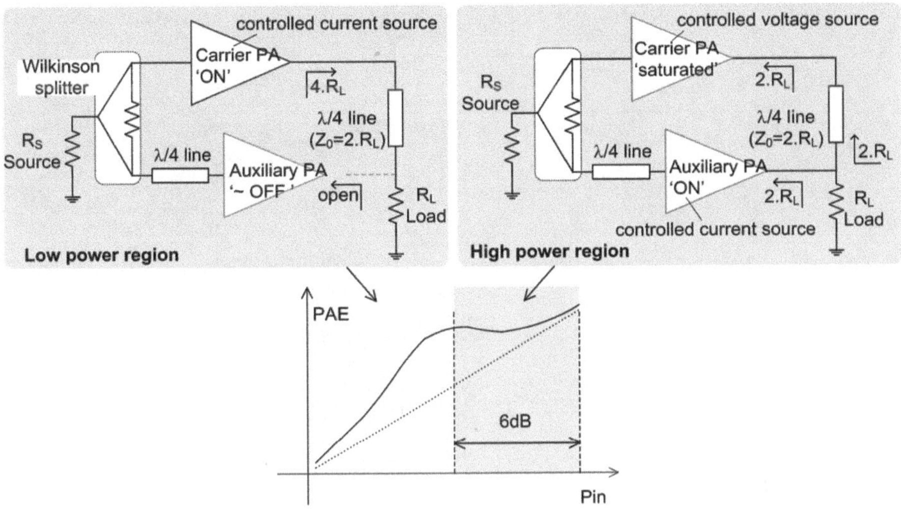

Fig. 1.20 Doherty amplifier architecture

feasibility to extend this value at the cost of higher complexity (Srirattana et al. 2005). In an analog manner with outphasing architecture, Doherty principle requires the use of quarter-wave lines that cannot be integrated on silicon for 2 GHz applications for die area reduction purpose. Therefore, it is generally not considered a good candidate for handset transmitters either. A Doherty-like alternative employing lumped elements was proposed in (Cha et al. 2003) though remaining area consuming.

1.2.3 Uplink-Compliant Efficiency Enhancement PA Architectures

1.2.3.1 Envelope Elimination and Restoration

The principle and context of *EER* were already partially presented in Section 1.1.2. It is based on the idea that magnitude and phase information can be separated and treated independently (Fig. 1.21). The phase of the *RF* modulated signal is probed by means of a limiter. The resulting signal is a *RF* square wave that is amplified by a high efficiency power stage, for example a switching class amplifier. Separate from this, the supply voltage V_{DD} of the power stage is made an image of the input envelope by means of a power detector and a supply modulator.

Switching class amplifiers (class D, E, and F/F^{-1} at high power level) feature the advantages that they leave the input phase information unchanged and that the output power is power supply-only driven:

$$P_{OUT} \propto V_{DD}^2 \tag{1.9}$$

where V_{DD} is the power supply voltage.

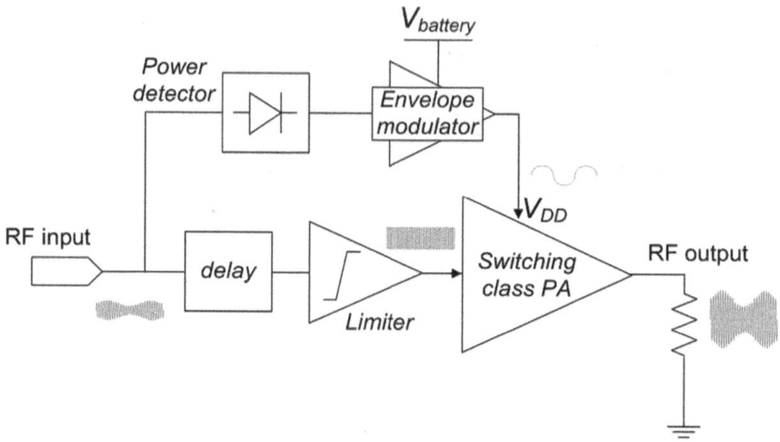

Fig. 1.21 Envelope elimination and restoration architecture

In this way, phase and magnitude are efficiently recombined. Meanwhile, *EER* also presents disadvantages, in particular its high sensitivity to magnitude/phase non synchronization. Indeed, 1 ns delay results in several excess *EVM* percents. Therefore, a phase shift with proper calibration must be integrated in the *RF* phase path. Nevertheless, the Magnitude/Phase synchronization is a challenging condition to fulfill over large channel bandwidths, which makes *EER* more complex to implement for future 3 G/4 G wireless standards. Another drawback lies in the efficiency drop due to the lossy response of the envelope modulator that proves to be a critical element. Indeed, the drain efficiency of a class E amplifier is theoretically as high as 80% and can be up to 65% in practice. But the overall *PAE* in a *EER* architecture never reaches such high value according to the following equation:

$$\eta_{overall}^{-1} \approx \eta_{PA}^{-1} + \eta_{envelope\_modulator}^{-1} \tag{1.10}$$

Moreover, even though the limiter, and the power detector can be easily implemented on-chip, the envelope modulator is difficult to integrate, die area consuming and may increase the bill of material. The following envelope modulator topology (Fig. 1.22) is relatively easy to integrate on silicon and requires no external device (Reynaert and Steyaert 2005).

Its drawback lies in its relatively poor efficiency at low power levels due to the drop-out voltage that results in a significant power loss:

$$P_{reg\_loss} \approx \left(V_{battery} - V_{DD}\right) \times I_{DD} \tag{1.11}$$

When associated with a class E power stage, this topology remains more efficient than class A amplification. Indeed, class E amplifiers present a current consumption that is power dependent thus mitigating the power dissipation in the regulator:

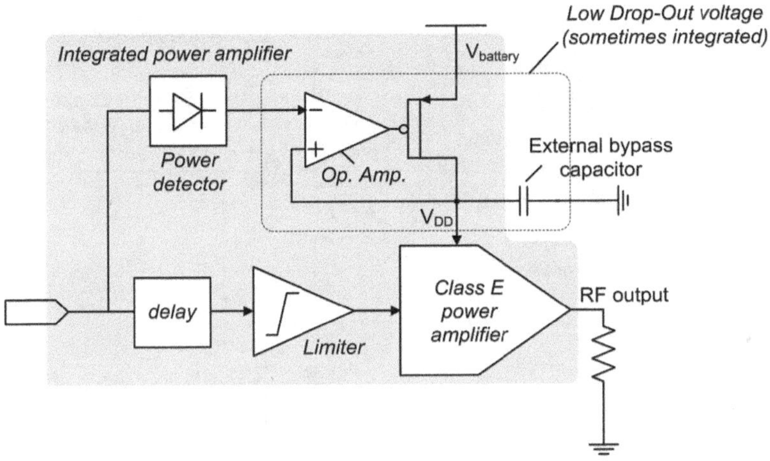

Fig. 1.22 *EER* using a drain voltage regulator

$$I_{DD}\,|_{class\ E} \propto P_{IN} \tag{1.12}$$

This topology is often used in *GSM/EDGE* power amplifier modules. The *EDGE* mode is only used when extended data rate is required, and the *GSM* mode is active for the majority of the time thus providing high efficiency since it operates at full power:

$$V_{DD_{GSM}} \approx V_{battery} - V_{drop\_out} \tag{1.13}$$

The most extensively developed *EER* topology in literature is based on DC–DC converters (Milosevic et al. 2003, Asbeck et al. 2005, Pinon et al. 2008, Walling et al. 2008). This ensures high efficiency at any power level but increases the bill of material. At least one high-value external choke inductor must be used to filter out the switching harmonics. The DC–DC converter is controlled by a square-wave voltage that can be provided either by a base-band *Digital Signal Processor* (*DSP*) or by an inner circuit. In many approaches, the envelope is first probed by a power detector and transformed into a voltage supply V_{DD} via a closed-loop that includes an envelope error amplifier, a loop filter, an *Analog to Digital Converter* (*ADC*) and a DC–DC converter. Figure 1.23 represents a typical illustration of this kind of topology.

In this example, the loop filter is included in the error amplifier, and a *Pulse Width Modulator* is used as an *ADC*. Such *PWM* block requires a triangle-wave voltage that can be generated from a clock. The loop bandwidth is determined by the loop filter, and the external choke inductor and bypass capacitor.

The clock frequency must be chosen consistently with the rise/fall time of the converter switches and is determined by the loop bandwidth and the targeted oversampling ratio.

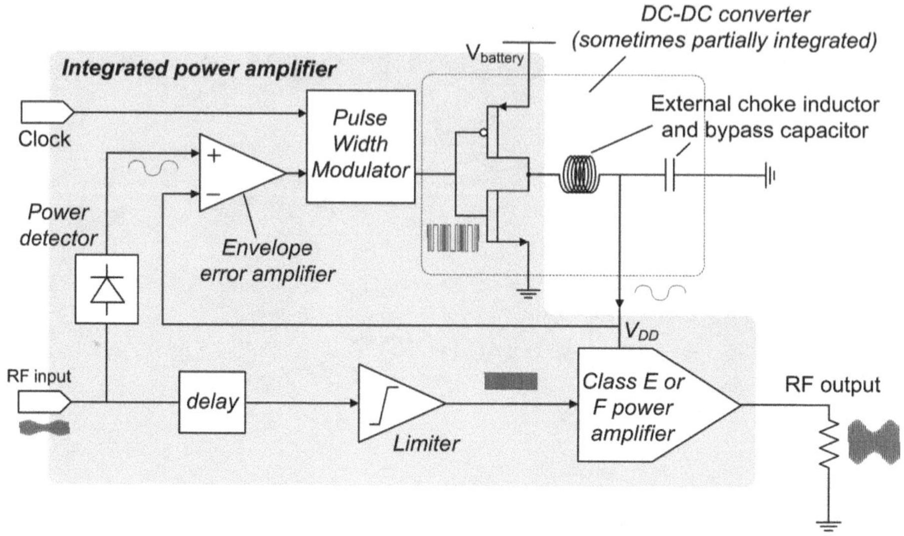

Fig. 1.23 *EER* using a pulse width modulator and a DC–DC converter

$$f_{clock} = 2f_{BW} \times OSR << \frac{1}{\tau_{fall/rise}} \tag{1.14}$$

where f_{clock}, f_{BW} are respectively the clock frequency and operation bandwidth, and *OSR* is the oversampling ratio, $\tau_{fall/rise}$ is the rise/fall time of the switches.

In some cases, the loop bandwidth may not prove to be sufficient, either because the channel bandwidth is too large, or because decreasing the rise/fall times results in excessive switch area consumption and/or efficiency degradation. To overcome this bandwidth limitation, the linear-assisted switched-modulator topology was proposed as depicted in Fig. 1.24 (Yousefzadeh et al. 2005, Kitchen et al. 2007, Blanken et al. 2008).

This is a dual mechanism that involves the previously presented envelope modulation principles. A high efficiency DC/DC converter supplies most of the necessary power, and a low-quiescent current voltage regulator intervenes as an assistant circuit when transition currents are pulled-down at too great a velocity by the *PA*.

According to (Presti et al. 2008), another issue of *EER* lies in its limited operation range. Indeed, below specific levels, the *PA* and the limiter become linear. In this transition region, the output envelope risks not being properly controlled by neither the input power nor by supply voltage V_{DD}, which increases distortion and requires compensation or predistortion techniques. Furthermore, at very low power level, *EER* no longer behaves as explained above but rather according to the envelope tracking principle that is described below.

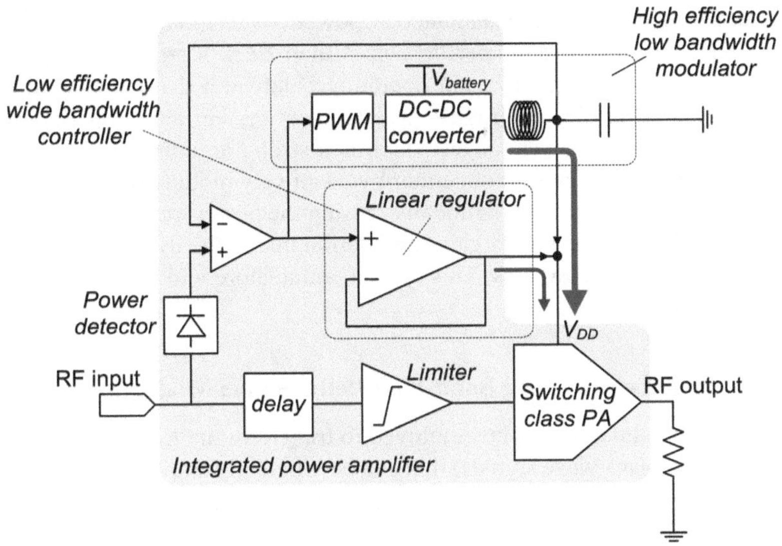

Fig. 1.24 *EER* using a linear-assisted switched modulator

1.2.3.2 Envelope Tracking

The *Envelope Tracking* (*ET*) principle does not imply any signal shaping/transformation but consists in configuring a linear power amplifier in the most appropriate conditions so as to simultaneously enhance both efficiency and linearity (Chen et al. 2008, Wang et al. 2008, Deltimple et al. 2005). The basic principle is depicted in Fig. 1.25. The power capability of a class A/AB amplifier is dynamically

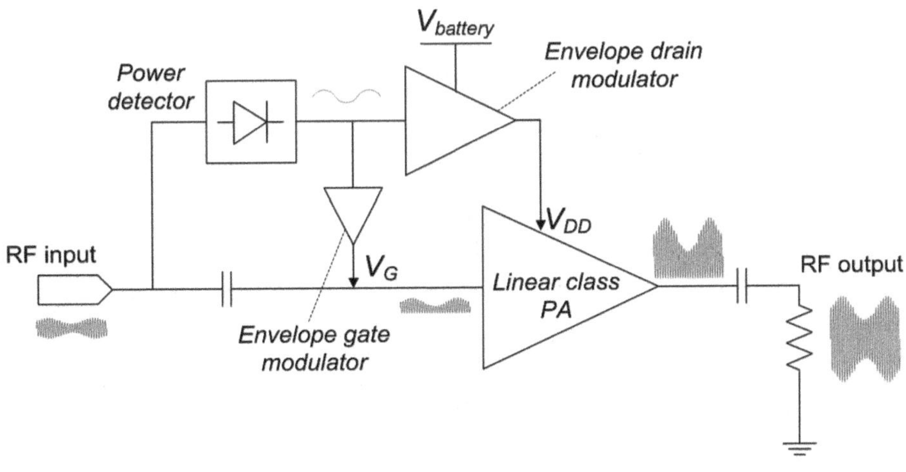

Fig. 1.25 Envelope tracking architecture

adapted as a function of the instantaneous power. Even though such principle is known to have lower efficiency performances than *EER*, it strongly alleviates the critical issues of synchronization and operation bandwidth that *EER* suffers from. In addition, this principle requires no predistortion system and operates continuously over a very wide power range, which is a strong argument in its favor. The envelope tracking can be carried out either by adaptively modulating the power supply (drain/collector *ET*) or by dynamically biasing the power transistor, or both. The drawback of *ET* lies in the power gain modulation due to the dynamic modification of the power stage characteristics. This type of architecture will be more extensively developed in Chapter 3.

1.2.3.3 *PA* Architecture Using Band-Pass Delta–Sigma Modulator

A Delta-Sigma modulator may be employed to transform the *RF* input signal into a digital data (i.e. square-wave signals) that can be efficiently amplified by a switching class amplifier, for example a class-S power stage (Fig. 1.26). Delta sigma modulation rejects the quantization noise outside the band of interest. The original signal is simply recovered thanks to the output band-pass filter that cancels the out-of-band spurious.

This principle is not limited by synchronization problems such as *LINC* or *EER*. No additional external passive device is required. The power range over which efficiency is maximized is determined by the dynamic range of the ΔΣ modulator and is much broader than that of Doherty or *EER*. Meanwhile, some issues arise with this topology. First, it tightens the out-of-band rejection specification of the output filter. Second, designing a band pass ΔΣ *ADC* working at a few gigahertz, and

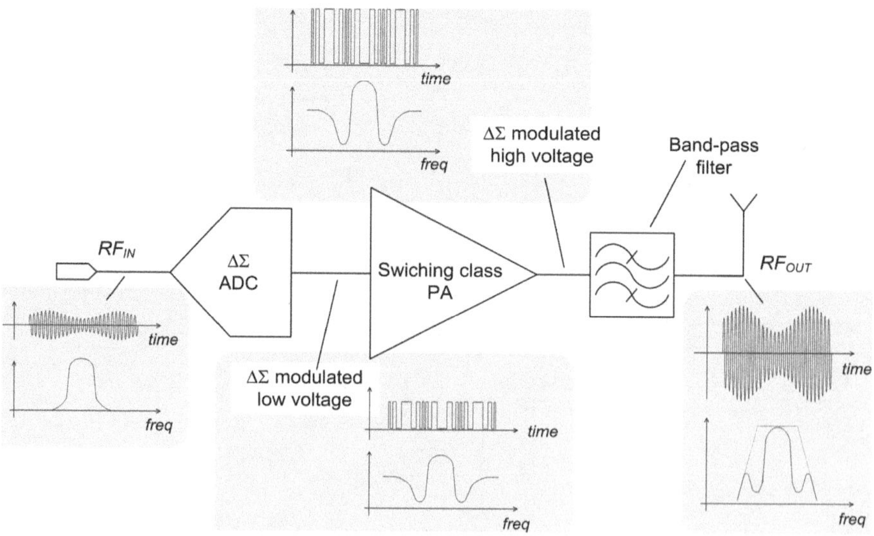

Fig. 1.26 Association of *PA* and band-pass ΔΣ modulator

featuring wide-channel bandwidth (20 MHz for *WLAN*), high dynamic range (74 dB for *WCDMA*) is a complex task and should require very high clock frequency. This is at the expense of high current consumption, which is in contradiction with an efficient transmitter. Most of the works presented to this date address the low band between 800 MHz and 1 GHz (Kelly et al. 2008, Ralph and Farrell 2007, Ketola et al. 2004, Uang et al. 2000). In order to extend the carrier frequency performance, a low power digital band-pass ΔΣ modulator was proposed in (Frappe et al. 2008). Nevertheless, this latter topology modulator was developed on nanoscale *CMOS* processes that are little compliant with high voltage *PA*. Therefore, power capabilities, high carrier frequency, wide channel bandwidths are difficult to combine simultaneously with this architecture.

1.2.4 Conclusion

From the previous review, *PA* architectures based on envelope modulator (*EER* for example) are very area-consuming and may be little compliant with complex multi-standard or *MIMO* applications for mobile handsets. The *Bill-of-material* and equipment area is a key factor for cellular phones manufacturers and must be limited to its strict minimum since the upper layer of printed circuit boards is essentially occupied by the user interface display/keyboard. This is also the reason why isolators are rarely used nowadays. The *PA* Class-S topology driven by a ΔΣ modulator are rarely considered due to bandwidth limitations and the resulting current consumption. The "ideal" *PA* may consist in combining several techniques such as drain(/collector), gate(/base) and/or load-line modulations while switching on/off stages in order to broaden power reconfigurability range. The point of view that will be developed throughout this book puts the stress on fully integrated solutions on silicon. and privileges low die area, "low consumption" architectures requiring neither external components, nor predistortion system. *Envelope Tracking* will be considered and declined in several manners. The development of various envelope detection and (discrete or continuous) actuation principles will be detailed in the prospect of *PA* dynamic reconfiguration and efficiency enhancement. Developed in cooperation with ST Microelectronics, the architectures that are described in this work try to put forward the advantages of 0.25 μm *BICMOS* technology and the interest of integrating high power devices with low power analog signal processing. From a technological point of view, the design of a power amplifier is ruled by severe trade-offs that will be discussed in the next section.

1.3 Technologies for Handset *PA* Design

1.3.1 Silicon Versus III/V

Despite the enhancement of *CMOS* technologies in the past few years, 60% of the power amplifier handset market is still dominated by *III/V* manufacturers such as

Table 1.7 Comparison of various technologies from *PA* design point of view

Technology	Price/watt	Power density	Linearity	Freq	Efficiency
SiGe HBT	Low	+	Medium	+	+
Si LDMOS	Low	–	++	–	Medium
GaAs pHEMT	Medium	Medium	++	++	+
InGaP HBT	High	+	+	++	+

Table 1.8 Comparative table of *III/V* and *SiGe* performances

		$BV_{DS} \times f_T$ (V.GHz)	$BV_{CE0} \times f_T$ (V.GHz)
ST Microelectronics B7RF	*HBT* HV	–	300
SiGe 0.25 μm *BICMOS*	*LDMOS* HV	450	–
Triquint TQHBT3 : *InGaP HBT*		–	560
Triquint TQP15 : 0.15 μm D-mode *GaAs pHEMT*		1050	–
RFMD *GaN HEMT* on *SiC*		2200	–

RFMD, Skyworks and Triquint (Zampardi 2008, Metzger et al. 2007). In practice, *GaAs pHEMT* or *InGaP HBT* are the most commonly used devices for the design of handset *PA* modules. These technologies present the best performances in terms of transition frequency, breakdown voltage, ruggedness, linearity, power density and turn-around time.

Table 1.7 summarizes the properties of currently employed technologies for *PA* design.

The figure of merit $f_T \times BV$ (transition frequency multiplied with the drain source breakdown voltage) is a useful criterion to evaluate the compliance of a technology with high power *RF* applications. Table 1.8 provides comparative data for various processes. Due to its low cost and high linearity, silicon *LDMOS* is chiefly employed in the market of *PAs* for base-stations. Technologies on *GaN* or *SiC* substrates also seem very promising since they have improved *RF* and thermal performances. Despite lower linearity, they are targeted to compete with silicon *LDMOS*, especially for *WIMAX* or *LTE* infrastructure deployment.

The previous discussion raises the issue of the role that *CMOS/BICMOS* manufacturers may play in the *PA* market in the next few years. Leading companies in the field of handset power amplifiers generally propose multi-mode multi-band *PA* modules in a *System-In-Package* assembly (*SIP*). Such miniaturized and complex *SIP* assemblies include one or several *GaAs/InGaP* power dice, *CMOS* controller *ICs* and integrated couplers/detectors. The trend will move towards even higher complexity in future modern wireless terminals. Indeed, still more numerous operation bands are being jointly addressed on a single module (*GSM*, *EDGE*, *WCDMA*, *LTE*, Bluetooth, *WIMAX*, amongst others), as well as more functionalities (power management, *MIMO*, amongst others).

These two requirements result in an increase of the necessary *SIP* laminated area, of the bill-of-material (high number of dice, passive filters and so on) and of cost. This may move in favor of single-die solutions and stresses the advantages of *BICMOS* technologies their relatively low cost, continued good performances, their ability to easily co-integrate power amplifiers, with power detectors, controller circuits, sensors (temperature probe for example) and any other analog/digital blocks on the same die; co-integration which is quite impractical with *GaAs* processes.

1.3.2 Presentation of ST Microelectronics BICMOS 0.25 μm Technology

1.3.2.1 Process Brief Overview

ST Microelectronics *BICMOS* 0.25 μm technology (*B7RF*) is a *SiGe* process and allows the full range of digital, analog, as well as radiofrequency applications to be covered. Despite the relatively high number of physical layers to be processed, its cost per square millimeter remains low compared to the most advanced *CMOS* counterparts. It features a good trade-off between integration capability, breakdown voltage, and *RF* performances due to its high substrate resistivity. It provides a thick top metal layer (Al or copper option) that enables the design of medium quality factor passive devices (inductor Q up to 10 at 2 GHz) and provides design ruggedness to electromigration (up to 5 mA/μm) at the cost of low parasitic capacitive effect. The active devices include *MOS* and *SiGe heterojunction bipolar transistors* (*HBT*) for high speed low noise applications, as well as laterally doped MOS (*LDMOS*) and high voltage *HBT* for power amplification. *LDMOS* devices are dedicated to high frequency applications and must be distinguished from *Drift MOS* that are geometrically similar (*LOCOS* inclusive) and that are devoted to power management *IO's* (low frequency and high voltage properties). Extra components such as Schottky diode for power detection are also available.

1.3.2.2 Laterally Doped *MOS* Transistors

Laterally Doped MOS (*LDMOS*), sometimes known as *Laterally Double Diffused MOS* are among the best solutions for high power *RF* applications and are extensively used in the field of base station power amplifiers. Their electrical behavior is fundamentally close to that of basic *MOS* but they present enhanced performance in terms of drain voltage capability (higher breakdown voltage) and consequently ruggedness to *VSWR* mismatch. Moreover, they present an asymmetric physical aspect and their layer stack is different in many ways. The cross-section of a STMicroelectronics single-sided halo *nLDMOS* (*B7RF*) is depicted in Fig. 1.27. Contrary to classic *MOS*, the (low doped) *pBody* region below the gate where the inversion layer is generated may present a non-uniform lateral doping. Source and *pBody* are locally connected by means of the salicide deposit.

In addition, a lightly doped drift region is implanted below the gate oxide nearby *pBody*. The structure in Fig. 1.27 is called a single-sided halo. Indeed, *pBody* and

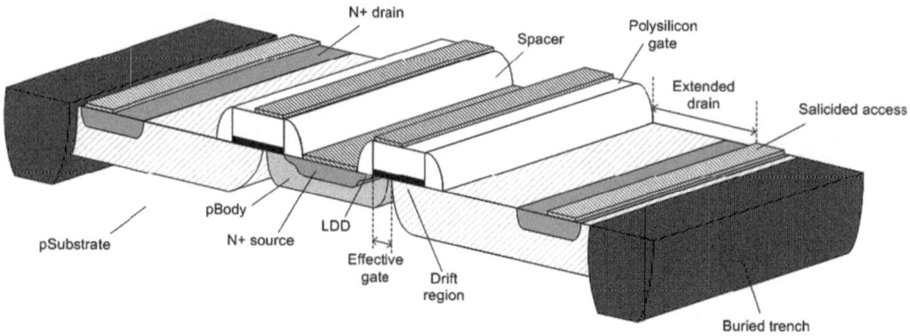

Fig. 1.27 *LDMOS* cross-section (ST Microelectronics dual-gate structure)

nDrift region are not directly abutted, but linked via a highly doped *pSubstrate* interface, in order to reduce punch-though effect (Knaipp et al. 2006). The drift region presents an unsalicided extension to the effective drain access in order to increase breakdown voltage (BV_{DSS}). Unfortunately, this is at the cost of reduced *RF* performances and additional drain resistance. Generally speaking, the doping distribution gradient along the gate and drain must be optimized in order to preserve a good transition frequency (f_T) while avoiding strong electric field peaks resulting in hot carrier injection. Channel engineering is therefore of major importance, and determines the two fundamental trade-offs that are implied in *LDMOS*: $BV_{DS} \times f_T$ and $BV_{DS} \times (R_{ON})^{-1}$. It was demonstrated that the performances of single-sided halo structures are significantly higher than that of uniformly doped channels (Nihar 2005). Depending on foundries, some technological alternatives are proposed and debated on. For example, it has been discussed that the insertion of a low doped buried N_{TUB} (embedding the *pBody*) increases parasitic drain/bulk capacitance, resulting in higher losses in the bulk and a decrease of efficiency. Further, protection structures at the gate/drain interface such as the insertion of a *LOCOS/STI* (*Shallow Trench Isolator*) are sometimes employed. It has been reported that the use of locos makes the removal of gate salicide unnecessary, which allows for the reduction of the gate access resistance (Muller et al. 2005). On the other hand, locos result in higher hot carrier injection and magnified kink effect below the threshold (Wang et al. 2008). Indeed, kink effect is strongly related to the oxidation/diffusion processing step (*OED*) and is of great concern for *LDMOS* that are based on double diffusion. As will be explained in Chapter 3, the principle of an adaptive power amplifier is based on the control of power transistors over a large bias range, ideally down to threshold. This means that the modelling discrepancies due to kink effect at low gate voltage level are highly undesirable for linearity performances. As *B7RF LDMOS* devices do not incorporate *LOCOS*, nor buried N_{TUB} layer, they remain a good candidate for our high speed, wide bias range application.

LDMOS modelling is significantly more complex than that of classic *MOSFET* and is generally addressed according to several methodologies:

- Empirical model
- Database model
- Sub-circuit-based model.

The latter presents the advantage to remain physically meaningful and combines accuracy and scalability. This method is used by many foundries, ST Microelectronics amongst others. Based on the literature, a generic sub-circuit-based *LDMOS* model may be represented by Fig. 1.28 (Pallotta 2006, Ankarcrona et Olsson 2002, Canepari et al. 2005).

It consists in several sub-*MOSFETs* built on *BSIM3v3* model, as well as npn/pnp bipolar transistors/diode, associated with additional resistances and capacitances. Physically speaking, the *MOSFET* referenced here as $M_{channel}$ represents the effective gate at the interface between *pBody* and oxide gate. In *B7RF* technology, its length is physically fixed to 0.2 μm. The *MOSFET* called $M_{pinch\text{-}off}$ is sometimes included to represent the effect of the lightly doped drift region below the gate, depicted by the *nLDD* region in Fig. 1.27. The *MOSFET* referenced here as $M_{drainext}$ represents the current dependent drift region resistance. C_{gd} is modeled by means of a non-linear *MOS* capacitor.

ST Microelectronics *SiGe* 0.25 μm technology provides two categories of *LDMOS* devices:

- High speed devices presenting medium breakdown voltage
- High voltage devices featuring medium transition frequency

Transition frequencies for both devices are depicted in Fig. 1.29.

It must be noted that transition frequency remains roughly constant over a wide range of gate bias condition. This is a very interesting property in the case of an adaptive *PA*, and should be considered when choosing the most appropriate device for the *PA* design.

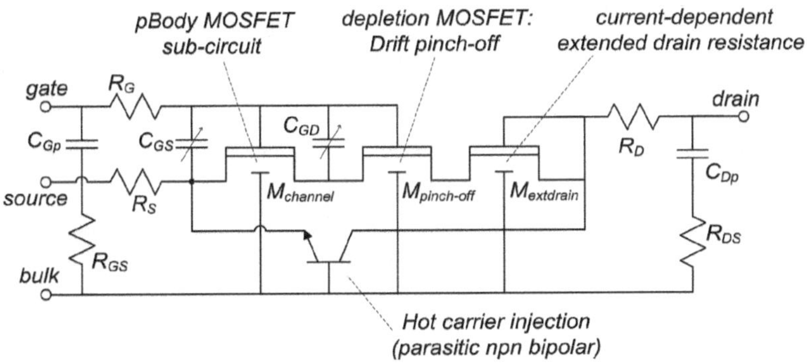

Fig. 1.28 Sub-circuit-based *LDMOS* model

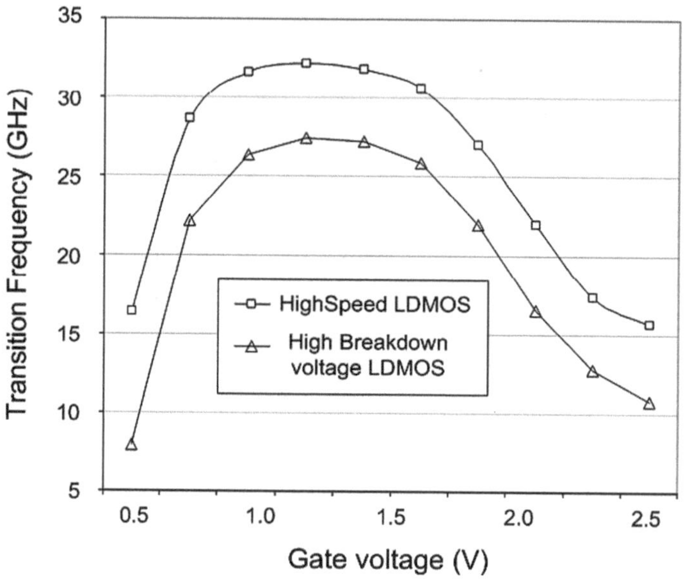

Fig. 1.29 *B7RF LDMOS* transition frequencies

The main characteristics of *B7RF LDMOS* are summarized in Table 1.9 for a 10 μm-wide device (Szelag et al. 2006). For comparison purpose, Table 1.10 lists the *LDMOS* figures of merit for various technologies.

When focusing on linearity, an important figure of merit lies in the transconductance non-linearity. Indeed, they play a dominant role in the *AM/AM* response of a

Table 1.9 ST microelectronics *B7RF LDMOS* characteristics

	2nd G *LDMOS*	2nd G high speed *LDMOS*
Max. transition frequency f_T (GHz)	28	32.5
Breakdown voltage B_{VDSS} (V) @I_{DD}=1 μA	16	9
Saturation drain current I_{dssat} (A.mm^{-1})	0.440	0.505
R_{ON} (Ω.mm)	2.8	2.05

Table 1.10 Comparative table of *LDMOS* silicon technologies for handset market

	$BV_{DS} \times f_T$ (V.GHz)	$BV_{DS} \times R^{-1}{}_{ON}$ (V.Ω $^{-1}$.mm^{-1})
STM *B7RF* (*SiGe* 0.25 μm *BICMOS*)	450 (L_D=0.5 μm)	6 (L_D=0.5 μm)
IHP (*SiGe* 0.25 μm *BICMOS*) (Nihar 2005)	414 (L_D=0.6 μm)/ 630 (L_D=1.4 μm)	2.73 (L_D=0.6 μm)/ 3.6 (L_D=1.4 μm)
Ericsson (0.35 μm *BICMOS*) (Bengtsson et al. 2003)	600 (L_D=1 μm)	1 (L_D=1 μm)

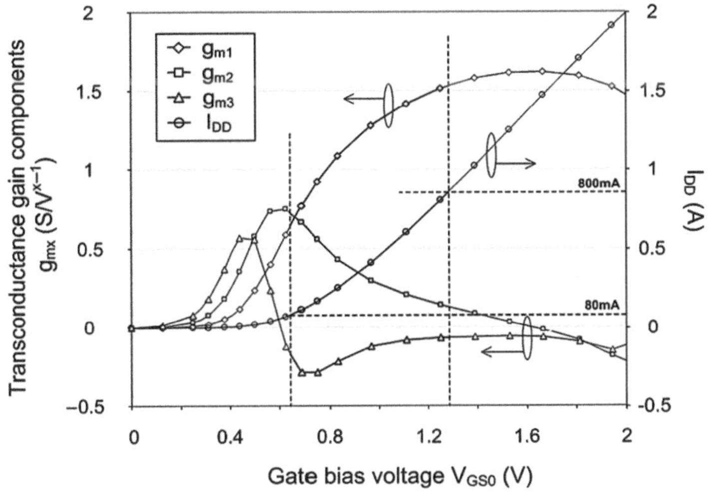

Fig. 1.30 I_{DD} and the first three transconductance components of a high BV_{DS}, 6.4 mm-*LDMOS*

PA, i.e. the extent to which the output *RF* magnitude is compressed (or magnified) as a function of the input *RF* magnitude. The first three order components are depicted in Fig. 1.30.

This curve is a fundamental element in the discussion over linearity in adaptive power amplifier design. The linear transconductance g_{m1} remains roughly constant at high bias level but quickly decreases for low bias level, i.e. in the bias region where the *PA* will operate most of the time (depicted by dark lines in Fig. 1.30). This results in non linear 2nd and 3rd order components whose magnitude becomes quite high when the gate voltage V_{GS} is below 0.8 V. Fortunately, in this bias range, g_{m2} and g_{m3} terms are opposite terms (in an analog manner to *MOSFET*). This interesting property can be advantageously employed when linearity needs to be optimized. Indeed, by a proper combination of *RF* and envelope harmonics, it becomes possible to make second and third order non-linear components compensate each other for reduced current consumption, thus resulting in an *AM/AM* inflection point. An adaptive bias power amplifier using this approach will be described in Chapter 3.

In an analog manner, 2nd and 3rd order non-linearities in *LDMOS* capacitances (especially C_{gs} and C_{gd}) can be characterized (Figs. 1.31 and 1.32). They play a major role in the *AM/PM* response of a power amplifier, i.e. the phase shift of the output *RF* carrier as a function of the input *RF* magnitude. In the particular case of *B7RF*, capacitances $C_{gs,1}$ and $C_{gd,1}$ present smooth and monotonous variations. Therefore, $C_{gs,2}$ and $C_{gd,2}$ present low and positive values. Third order capacitances $C_{gs,3}$ and $C_{gd,3}$ present a negative peak just above the threshold. Due to these properties, the various contributions to phase distortion may combine destructively. The response *AM/PM* vs. power may therefore present an inflection point (see Chapter 3 for details) and is better controlled over a wide bias/power range than with a bipolar device.

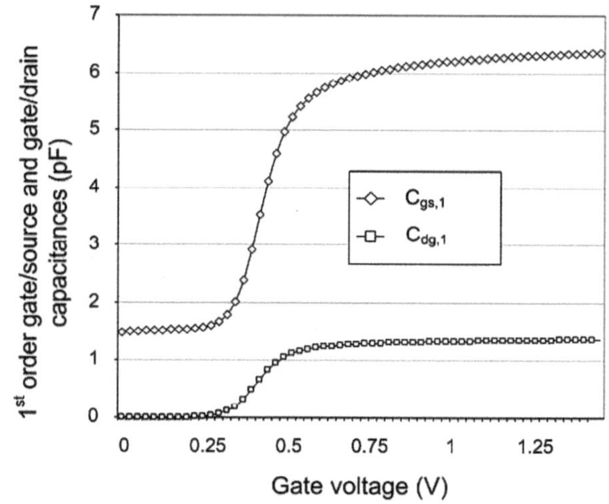

Fig. 1.31 Linear gate/source and gate/drain capacitances of a high BV_{DS} 6.4 mm-*LDMOS*

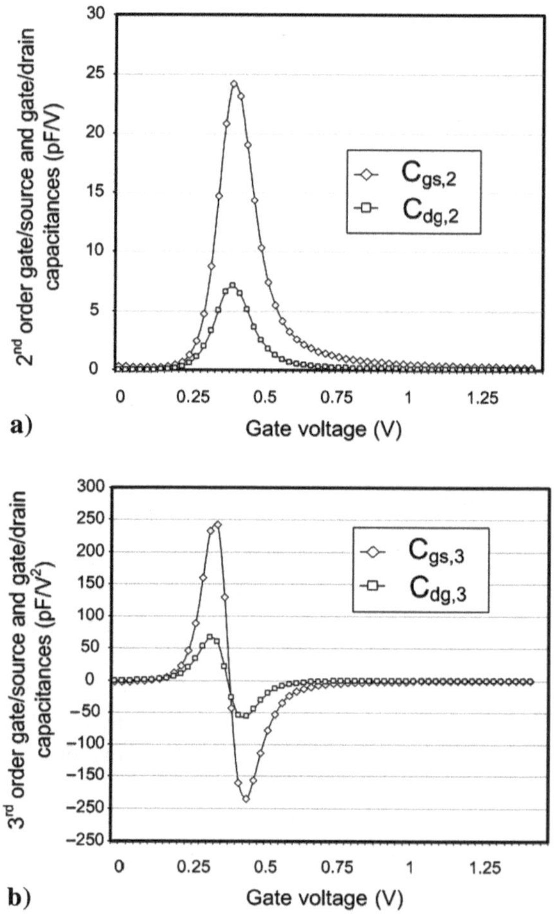

Fig. 1.32 (**a**) 2nd and (**b**) 3rd order characteristic capacitances of a high BV_{DS} 6.4 mm-*LDMOS*

An interesting property of *LDMOS* is that drain current I_{DS} decreases with temperature which ensures a stable thermal behavior. Indeed, the inversion voltage Φ_D and consequently the threshold voltage increase with temperature:

$$\frac{\partial V_t}{\partial T} = \frac{\partial}{\partial T}\left(V_{FB} + \Phi_D + K_B \cdot \sqrt{\Phi_D}\right) > 0 \tag{1.15}$$

where V_{FB} is the extraction potential (that allows flattening bands at metal/semiconductor interface), Φ_D is the inversion voltage.

Meanwhile, dynamic temperature variation can give rise to unexpected distortion, especially for adaptive bias *PA*. Indeed, the dynamic temperature swing resulting from the *RF* envelope behaves as a memory effect and is fed back to the *PA* itself with a frequency-dependent delay, thus modulating its characteristics. Therefore, accurate modelling of self-heating must be carried out. STMicroelectronics *LDMOS* model includes self-heating by means of a R_{th}, C_{th} parallel network (namely the thermal resistance and capacitance of the *LDMOS* structure on its own). Practically, the thermal sub-network of the whole environment must also be taken into account (i.e. substrate, packaging, interconnection, air interface...). This results in a substantial increase in thermal resistance whose weight can no longer be neglected in the case of a wide *LDMOS* matrix.

When high drain voltage are applied to a *LDMOS*, strong increase of drain current I_{DS} is observed due to non linear effects such as punch-through, and hot carrier injection. Punch-through is a somewhat soft effect and is related to the proximity of source and drain space charge regions at high drain voltage. Hot carrier injection is a much more sudden and sometimes destructive phenomenon that occurs when impact ionization generates electron-hole pairs in the vicinity of gate dielectric and when carriers get trapped. Hot carrier generally impacts the threshold voltage or the transconductance of the device and is the root of fast or slow memory effects.

Several hot carrier injection mechanisms can be distinguished, among others *Channel Hot-Electron* injection (*CHE*) and *Drain Avalanche Hot Carrier* injection (*DACH*). The latter is depicted in Fig. 1.33 and can be understood as a two-step phenomenon according to (Skotnicki and Bœuf 2003). When drain voltage increases, I_{BODY} (the leakage current in the *pBody*) increases whereby *pBody* self-biases in the channel vicinity (①). The threshold voltage V_t therefore decreases which makes the drain current increase. Then, a parasitic npn bipolar transistor {*nDrift/pBody/nSource*} switches on (②) and generates an avalanche drain current.

1.3.2.3 Heterojunction Bipolar Transistors

In ST Microelectronics 0.25 μm *SiGe BICMOS* technology (*B7RF*), the modelling of *Heterojunction Bipolar Transistors (HBT)* is based on *HICUM*. This model is particularly interesting in high frequency *RF* applications requiring high collector current density. This compact and scalable model presents much higher accuracy than the *Modified Spice Gummel-Poon* model and takes into account self-heating.

The frequency transition and current gain β_{AC} of a typical high voltage *HBT* are depicted in Fig. 1.34. The impact of the *Kirk* effect is visible on the right-hand side in

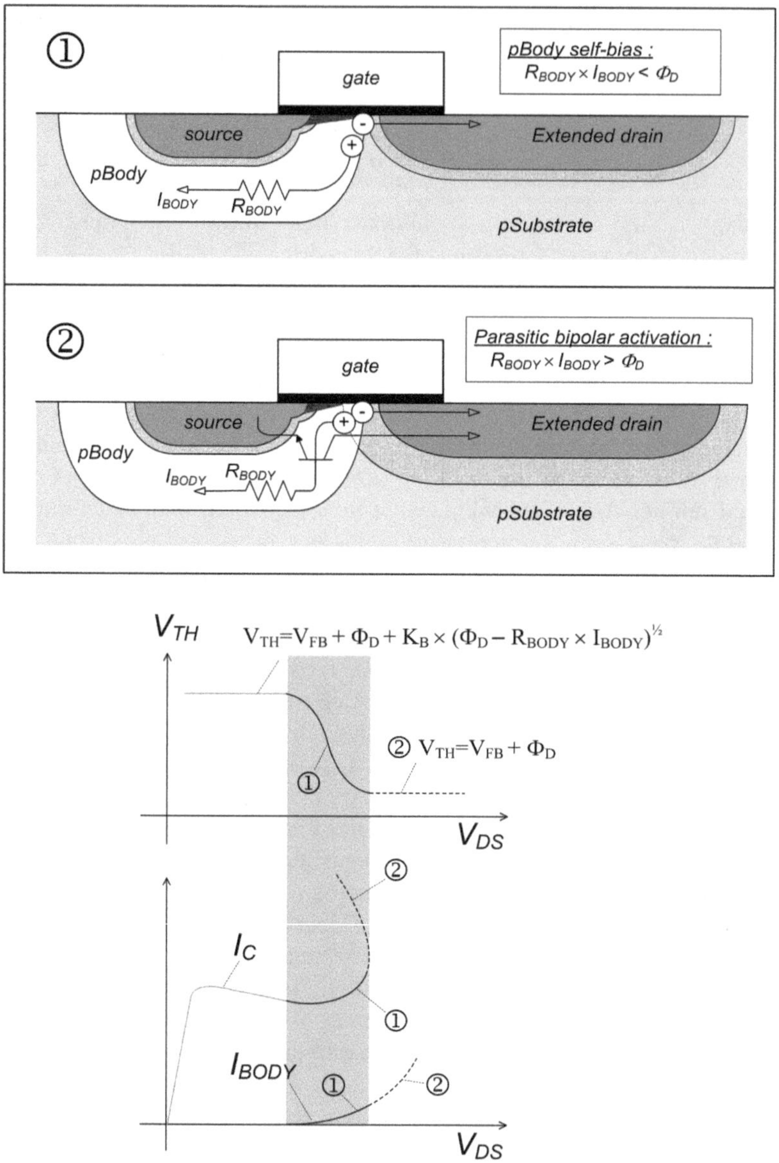

Fig. 1.33 Illustration of drain avalanche hot carrier injection in a *LDMOS*

grey and results in strong f_T and β gain collapse at high bias level. From a physical point of view, this effect is due to an increase of the base depth (base/collector junction is moved downwards) that opposes to carrier injection.

HBT transistors generally cannot be used on their own and require the use of ballast resistors at their emitter and base ports. The design of such degenerating components is dictated by complex trade-offs that are summarized in Fig. 1.35.

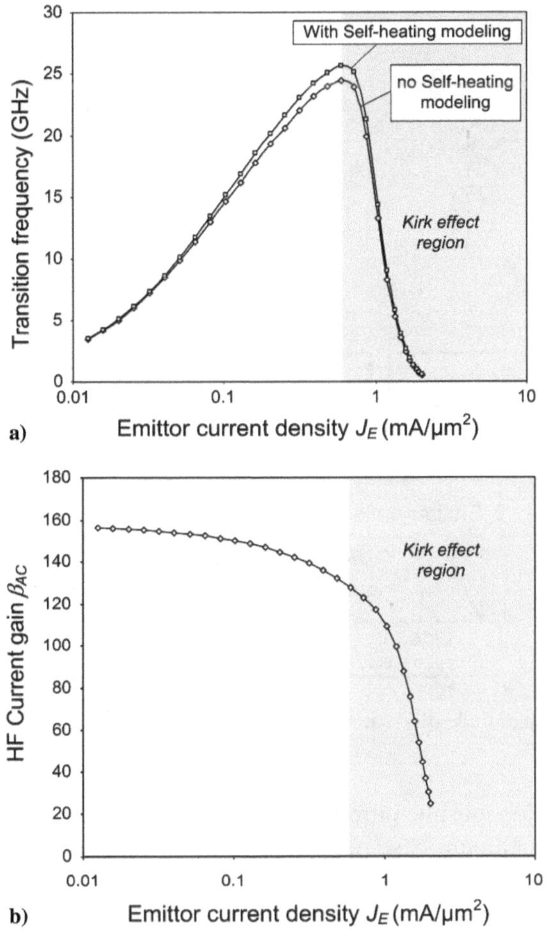

Fig. 1.34 *B7RF* high voltage *HBT* performances: (**a**) transition frequency, (**b**) AC current gain

In practice, the base ballast resistance is distributed as shunt and series elements. Schematic (a) in Fig. 1.35 represents the case of shunt base degeneration. Schematic (b) in Fig. 1.35 represents the case of series base degeneration. Both configurations have different properties. In Fig. 1.35, the signs "−" and "+" must be respectively understood as a decrease and an increase. When two signs are separated by a "/", the left sign refers to configuration (a) and the right one to the configuration (b).

Emitter degeneration is always an absolute necessity in order to thermally stabilize and linearize the device. Nevertheless, its value is limited in practice because of its strongly negative impact on power gain and efficiency. A practical rule of thumb to help preserve efficiency is that the voltage drop ($I_C \times R_E$) should not exceed twice the thermal voltage (∼50 mV at room temperature).

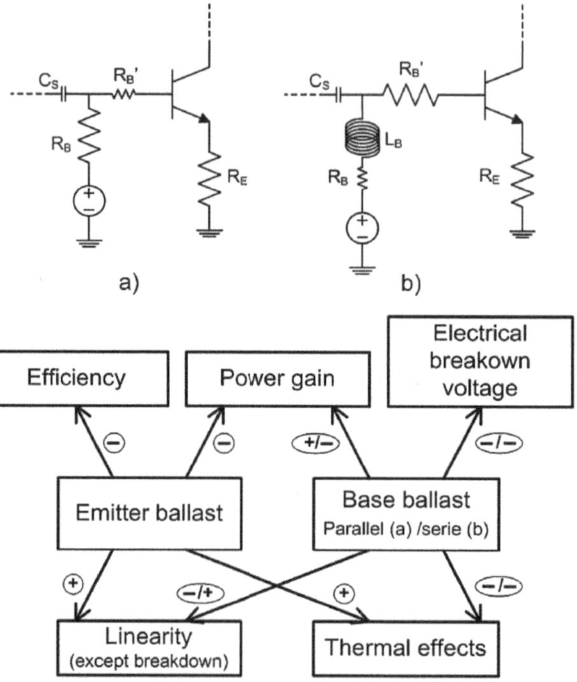

Fig. 1.35 Description of trade-offs that rule the design of ballast resistors

Base degeneration must be introduced as a complementary thermal ballast (Chu et al. 2000). The optimum base resistance value is complex to determine since it is dictated by the trade-off *electrical breakdown/secondary thermal breakdown*. The electrical breakdown voltage decreases from BV_{CB0} (12 V) to BV_{CE0} (5.7 V) when base resistance increases. Contrarily, a breakdown effect due to thermal heating arises from the lack of base degeneration at low frequency (in the base-band frequency domain) and may result in severe non-linear response (see Figs. 1.36 and 1.37).

In addition, linearity performance is also ruled by the impedance values at the transistor ports in the base-band frequency range and at twice the carrier frequency ($2f_C$). These are known as the *out-of-band terminations*. It is vital this notion be taken into account when designing a *PA* whatever the active device might be (*MOS*, *LDMOS* or *HBT*). Second order nonlinear harmonics are enhanced or mitigated according to the base resistance at 0 Hz and $2f_C$ (Deng et al. 2004, Aparin and Larson 2003). According to (Spirito et al. 2005), the optimum base terminations that allow minimizing distortion for a *HBT* are:

$$Z_{B,DC} = Z_{B,2f} = \frac{\beta}{2 \cdot g_m} \tag{1.16}$$

where β is the *HBT* current gain, and g_m is the transconductance gain.

For more flexibility, (Deltimple et al. 2005) proposed a bias circuit topology for *HBT PAs* that allows reconfiguring bias current and base impedance in an independent manner. A compromise between configurations (a) and (b) generally needs to be found in order to maximize both electrical and thermal breakdown voltages while minimizing the base-band 2nd order harmonics.

The thermal behavior of a *HBT* is mainly determined by the thermal dependences of the current gain β and of the *SiGe* band-gap E_g as shown by the approximated (1.17).

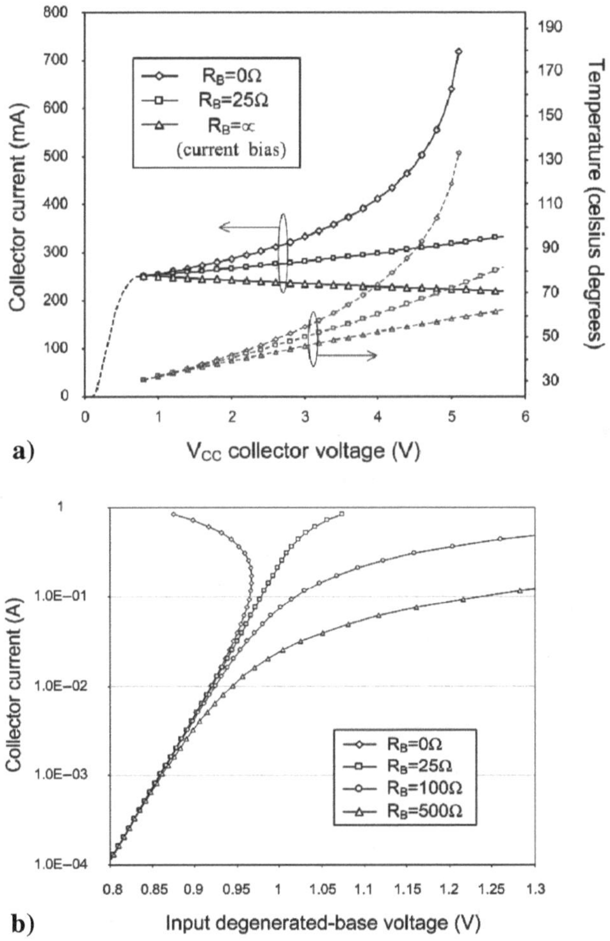

Fig. 1.36 Illustration of base ballast impact on self-heating: (**a**) I_C vs. V_{CE} characteristic, (**b**) I_C vs. V_B characteristic (Flyback effect)

Fig. 1.37 Illustration of collector current collapse in multi-finger bipolar transistors

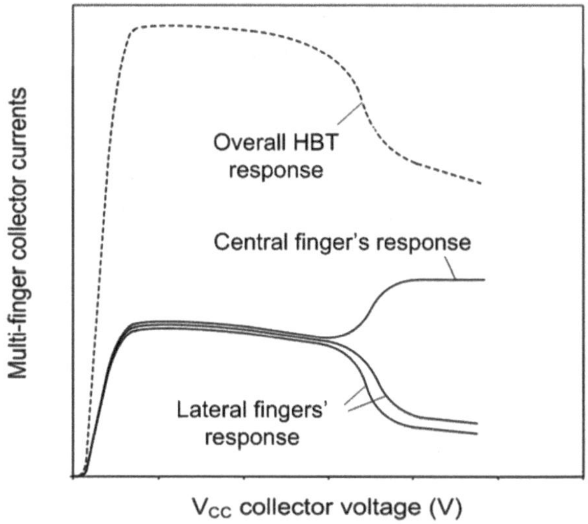

$$I_C(T) = I_{S0}(T) \cdot \left(1 + \frac{V_{CC}}{V_{early}}\right) \cdot \exp\left(\frac{V_{IN} - I_C(T) \cdot \left(R_E + \frac{R_B}{\beta(T)}\right)}{n_C \cdot U_t(T)} - \frac{E_g(T)}{U_t(T)}\right) \tag{1.17}$$

where I_{S0} is the inverse current, V_{CC} is the power supply, V_{early} is the Early voltage, V_{IN} is the input base bias voltage, R_E and R_B are respectively the emitter and base ballast resistors, n_C is a correcting factor, k is the Boltzmann constant, q the electron charge, U_t is the thermal voltage $U_t = \frac{kT}{q}$.

The band-gap E_g and the current gain β both decrease with temperature as shown by (1.18) and (1.19).

$$E_g = E_{g0} + \frac{A_{EG} \cdot T^2}{B_{EG} + T} \tag{1.18}$$

$$\beta = \beta\left(T_0\right) \cdot \left(\frac{T}{T_0}\right)^{X_{TB}} \tag{1.19}$$

where E_{g0} is the intrinsic band-gap energy, $A_{EG} < 0$, $B_{EG} > 0$, and $X_{TB} < 0$.

For low (resp. high) base degeneration, the influence of β is mitigated (resp. magnified) and the slope of I_C vs. V_{CC} increases (resp. decreases) as detailed in (Sinnesbichler and Olbrich 1999). The thermal impedance that is included in the *HICUM* model is underestimated since it does not take into account the *PA* environment. In order to properly design the ballast elements and parry the lack of accuracy in the thermal network, numerical computations (*Maple*) of the collector current were carried out according to the method described in (Tseng et al. 2000), assuming a large *HBT* ($L_E = 600$ μm) and a thermal impedance as high as 30 Ω/W (higher than the intrinsic *HBT*). Figure 1.36a clearly highlights the thermal breakdown for low base resistance and high drain voltage values ($V_{CC} > 4$ V) that was diagnosed in (Rinaldi et al. 2006). The script that was employed is given in Annex A, Table A.1. The technological values were modified for confidentiality purpose but remain in a realistic order of magnitude. In an analog manner, for a current biased *HBT*, the collector current as a function of the base-degenerated input voltage was plotted thus revealing undesirable *flyback* for low base resistance as depicted in Fig. 1.36b. The script that was employed is given in Annex A, Table A.2.

Flyback thermal breakdown can cause strong distortion and should be avoided by an appropriate ballast design. These methods allow the minimum base/emitter resistance to be determined thus preventing thermal distortion and/or runaway providing the thermal impedance is roughly known. Another thermal issue lies in the non homogeneous distribution of base current among emitter fingers (not taken into account in the previous calculus). This is particularly true in the case of multi-finger *HBT*, i.e. when several emitters share a unique buried layer. The use of this kind of configuration means both the die area and the overall collector/substrate capacitance can be reduced. The drawback is a collector current collapse at high voltage level (Zhu et al. 2002), whereby linearity is degraded as depicted by the 3-finger *HBT* in Fig. 1.37.

1.3.2.4 Discussion About Silicon Active Devices

As a manner of conclusion on active devices, Heterojunction Bipolar Transistors present better performances than *LDMOS* in terms of current/power density and die area consumption. An application using switched *HBT* devices with fixed bias will be described in Chapter 2. Meanwhile, *LDMOS* will be chiefly employed throughout this document, and especially in Chapter 3, which describes an adaptive bias mechanism. When power gain is considered, *HBT* apparent superiority is mitigated by their need of accurate emitter/base ballasts. Even if most radiofrequency circuits are little sensitive to self-heating because of the relatively low thermal cut-off frequency (generally in the order of megahertz, i.e. much lower than the *RF* carrier frequency that is addressed in wireless applications), the situation is quite different for adaptive power amplifiers. In this case, dynamic temperature variations due to the *RF* envelope swing become of great concern. This is a decisive argument in favor of *LDMOS* that offer better warranties of thermal stability than *HBT* transistors for adaptive bias circuits.

1.3.3 PA Protection Against VSWR Variations

In parallel with efficiency enhancement techniques, it is often recommended to implement protection circuits around power amplifier cores. For cost purpose, phone manufacturers preferably remove isolators from wireless equipments. Consequently, power amplifiers must be designed so as to withstand high antenna *VSWR* levels up to 10 or more. To this end, technological solutions are sometimes found to be insufficient and compensation design techniques are required to extend the *VSWR* range

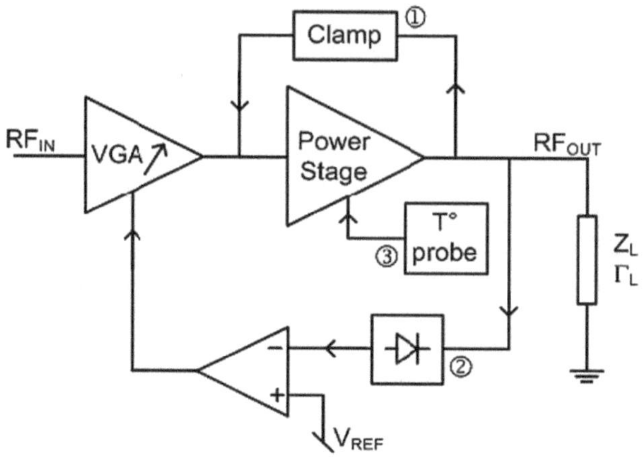

Fig. 1.38 *PA* protection techniques against *VSWR* variations

over which a power amplifier can safely operate. Moreover, such protections circuitry must be fast enough to cope with sudden over-voltage, over-current and/or over-temperature transients. High-bandwidth closed-loop topologies are therefore privileged the majority of the time. The general synoptic of a close-loop *VSWR*-protection architecture is simplified in Fig. 1.38. When an over-voltage arises at the load (i.e. when Γ_L presents a positive real part and $\Re(Z_L)$ is high), the power stage can be directly clamped by an active feedback loop ① (Yamamoto et al. 2000), or the voltage envelope peaks are detected and compared with a reference voltage thus reducing the gain of the *VGA* prior to the power stage ② and consequently the output voltage swing (Scuderi et al. 2004, van Bezooijen et al. 2007, Karoui and Parra 2008, Carrara et al. 2008).

When an over-current arises in the power device (i.e. when Γ_L presents a negative real part and $\Re(Z_L)$ is low), a temperature sensor probes the resulting thermal shift and dynamically actuates on the power stage bias current ③ (van Bezooijen et al. 2007). Over-voltage detection is generally implemented by sensing the sharp avalanche current through a blocked power device. One of the challenges in the design of protection circuits is that they must leave the *PA* performances unchanged in nominal load conditions, i.e. when $VSWR = 1{:}1$.

1.3.4 Presentation of ST Microelectronics Integrated PAssive Device (IPAD) Technology

Passive devices on silicon generally present low quality factors and are poor candidates for the design of efficient output matching networks due to their significant losses. This can be overcome by employing a low-loss passive technology. ST Microelectronics *IPAD RLC06* is a passive process on glass substrate that combines high *RF* performances and a high level of integration. This 3-metal-layer process

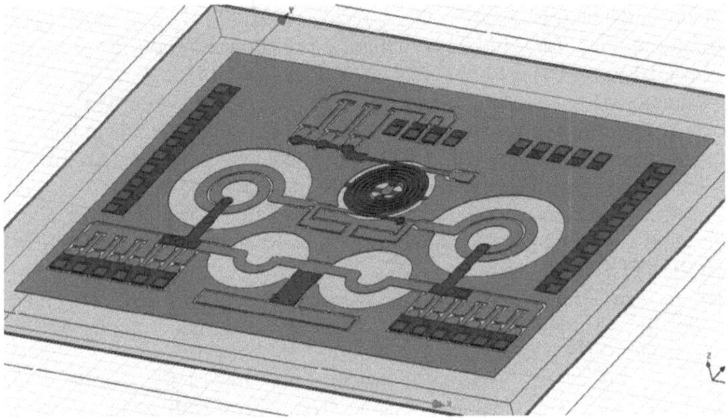

Fig. 1.39 3D view of a high-Q three port matching network on glass (ST *IPD* technology)

supplies a wide range of lumped and distributed passive devices, i.e. inductors, capacitors, baluns, as well as microstrip lines, coplanar waveguides, and suspended waveguides. Integrated inductors present improved quality factors as high as $Q = 50$ at 2 GHz and the capacitance density is up to 5 nF/mm^2. This technology allows for the implementation of low area high quality passive circuits, in combination with *ESD* protection zener diodes. *IPAD RLC06* is compliant with wire-bonded and flip-chipped assemblies and is a good candidate for *SIP* applications. The range of its applications is very broad, from wide-band *EMI* (electromagnetic interference) suppression filters for high data rate integrated circuits (such as displays or cameras) to *RF* matching networks or power combiners (Flament et al. 2008) in 3G handsets. An example of wide-band output matching network for a *PA* module will be presented in Chapter 3 (see its 3-dimensional view in Fig. 1.39). Due to confidentiality, the details and layer stack of *IPD* technology cannot be provided.

References

3GPP (1999) http://www.3gpp.org. Accessed 15 Dec 2010

3GPP (2006) R1-060023, Motorola Cubic Metric in 3GPP-LTE

Ankarcrona J, Olsson J (2002) Sub-circuit based SPICE model for high voltage LDMOS transistors. Phys Scr T101:7–9. doi:10.1238/Physica.Topical.101a00007

Aoki I, Kee S, Magoon R, Aparicio R, Bohn F, Zachan J, Hatcher G, McClymont D, Hajimiri A (2008) A fully integrated quad-band GSM/GPRS CMOS power amplifier. J Solid-State Circuits, 43(12):2747–2758. doi:10.1109/ISSCC.2008.4523311

Aparin V, Larson LE (2003) Linearization of monolithic LNAs using low-frequency low-impedance input termination. Proc Eur Solid-State Circuits Conf, Estoril, Portugal. pp 137–140. doi: 10.1109/ESSCIRC.2003.1257091

Apel T, Henderson T, Tang YL, Berger O (2008) Efficient three-state WCDMA PA integrated with high-performance BiHEMT HBT/E-D pHEMT process. Proc Radio Freq Integr Circuits Symp, Atlanta, USA. pp 149–152. doi:10.1109/RFIC.2008.4561406

Asbeck P, Larson L, Kimball D, Zhao Y, Wang F, Qiao D (2005) High dynamic range, high efficiency power amplifiers for wireless communications. Proc Bipolar/BiCMOS Technol Meet, Santa Barbara, USA. pp 103–107. doi:10.1109/BIPOL.2005.1555210

Bakalski W, Zannoth M, Asam M, Thomann W, Kapfelsperger B, Pfann P, Berkner J, Hepp C, Steltenpohl A, Österreicher W, Rampf E (2008) A load-insensitive quad-band GSM/EDGE SiGeC-bipolar power amplifier with a highly efficient low power mode. Proc Radio Week Symp, Orlando, USA. pp 203–206. doi:10.1109/RWS.2008.4463464

Bengtsson O, Litwin A, Olsson J (2003) Small-signal and power evaluation of novel BiCMOS-compatible short-channel LDMOS technology. Trans Microw Theory Tech 51(3):1052–1056. doi:10.1109/TMTT.2003.808697

van Bezooijen A, van Straten F, Mahmoudi R, van Roermund AHM (2007) Power amplifier protection by adaptive output power control. J Solid-State Circuits 42(9):1834–1841. doi:10.1109/JSSC.2007.900783

Birafane A, Kouki AB (2005) Phase-only predistortion for LINC amplifiers with chireix-outphasing combiners. Trans Microw Theory Tech 53(6):2240–2250. doi:10.1109/TMTT.2005.848748

Blanken PG, Karadi R, Bergveld HJ (2008) A 50 MHz bandwidth multi-mode PA supply modulator for GSM, EDGE and UMTS application. Proc Radio Freq Integr Circuits Symp, Atlanta, USA. pp 401–404. doi:10.1109/RFIC.2008.4561463

Canepari A, Bertrand G, Giry A, Minondo M, Blanchet F, Jaouen H, Reynard B, Jourdan N, Chante JP (2005) LDMOS modelling for analog and RF circuit design. Proc Eur Solid-State Device Res Conf, Grenoble, France. pp 469–472. doi:10.1109/ESSDER.2005.1546686

Carrara F, Presti CD, Scuderi A, Santagati C, Palmisano G (2008) A methodology for fast VSWR protection implemented in a monolithic 3 W 55% PAE RF CMOS power amplifier. J Solid-State Circuits 43(9):2057–2066. doi:10.1109/JSSC.2008.2001894

Cha J, Yang Y, Shin B, Kim B (2003) An adaptive bias controlled power amplifier with a load-modulated combining scheme for high efficiency and linearity. IEEE Microw Theory Tech Symp Dig, Philadelphia, USA. 1:81–84

Chauhan YS, Gillon R, Declercq M, Ionescu AM (2007) Impact of lateral nonuniform doping and hot carrier injection on capacitance behavior of high voltage MOSFETs. Proc Eur Solid-State Device Res Conf, Munich, Germany. pp 426–429. doi:10.1109/ESSDERC.2007.4430969

Chen CT, Li CJ, Horng TS, Jau JK, Li JY (2008) Design and linearization of class-E power amplifier for non-constant envelope modulation. Proc Radio Freq Integr Circuits Symp, Atlanta, USA. pp 145–148. doi:10.1109/RFIC.2008.4561405

Chireix H (1935) High power outphasing modulation. Proc Inst Radio Eng 23(11):1370–1392

Chu CY, Sheu JJ, Li GP (1999) Effects of ballast resistors on power and ESD performance in AlGaAs/GaAs heterojunction bipolar transistors. CS ManTech Conf Dig, Vancouver, Canada. pp 185–188

Cuoco V, van der Heijden MP, Pelk M, de Vreede LCN (2002) Experimental verification of the smoothie database model for third and fifth order intermodulation distortion. Proc Eur Solid-State Device Res Conf, Florence, Italy. pp 635–638

Cuoco V, Yanson O, Hammes P, Spirito M, de Vreede LCN, Steenwijk AV, Versleijen M, Neo WCE, Jos HFF, Burghartz JN (2004) Large signal verification of the circuit-oriented smoothie database model for LDMOS devices. Proc Eur Microw Conf, Amsterdam, The Netherlands. pp 217–220

Deltimple N, Kerhervé E, Belot D, Deval Y, Jarry P (2005) A SiGe controlled-class power amplifier applied to reconfigurable mobile systems. Proc Eur Microw Conf, Paris, France. doi:10.1109/EUMC.2005.1608892

Deng J, Gudem P, Larson LE, Asbeck PM (2004) A high efficiency SiGe BiCMOS WCDMA power amplifier with dynamic current biasing for improved average efficiency. Proc Radio Freq Integr Circuits Symp, Fort Worth, USA. pp 361–364. doi:10.1109/RFIC.2004.1320622

Doherty WH (1936) A new high efficiency power amplifier for modulated waves. Proc Inst Radio Eng 24(9):1163–1182

Dragon C, Brakensiek W, Burdeaux D, Burger W, Funk G, Hurst M, Rice D (2003) 200W push–pull & 110 W single-ended high performance RFLDMOS transistors for WCDMA basestation applications. Microw Theory Tech Symp Dig, Philadelphia, USA. 1:69–72

Flament A, Frappe A, Kaiser A, Stefanelli B, Cathelin A, Ezzeddine H (2008) A 1.2 GHz semi-digital reconfigurable FIR bandpass filter with passive power combiner. Proc Eur Solid-State Circuits Conf, Edimburgh, UK. pp 418–421. doi:10.1109/ESSCIRC.2008.4681881

Frappe A, Stefanelli B, Flament A, Kaiser A, Cathelin A (2008) A digital $\Delta\Sigma$ RF signal generator for mobile communication transmitters in 90 nm CMOS. ProcRadio Freq Integr Circuits Symp, Atlanta, USA. pp 13–16. doi:10.1109/RFIC.2008.4561375

Gajadharsing JR (2003) Low distortion RF-LDMOS power transistor for wireless communications base station applications. Microw Theory Tech Symp Dig, Philadelphia, USA. 3:1563–1566

Hadjichristos A, Walukas J, Klemmer N, Suter W, Justice S, Uppathil S, Scott G (2004) A highly integrated quad band low EVM polar modulation transmitter for GSM/EDGE applications. Proc Cust Integr Circuits Conf, Orlando, USA. pp 565–568

Hamedi-Hagh S, Salama CAT (2004) CMOS Wireless Phase-Shifted Transmitter. J Solid-State Circuits 39(8):1241–1252. doi:10.1109/JSSC.2004.831786

Hammes PCA, Jos HFF, van Rijs F, Theeuwen SJCH, Vennema K (2004) High efficiency, high power WCDMA LDMOS transistors for base stations. Microw J, vol 47, 4:94–101

Holma H, Toskala A (2007) WCDMA for UMTS: HSPA evolution and LTE, 4th edn. Wiley, New York, NY. ISBN: 978–0–470–68646–1

Jayaraman A, Chen PF, Hanington G, Larson L, Asbeck P (1998) Linear high efficiency microwave power amplifiers using bandpass delta–sigma modulators. Microw Guid Wave Lett 8(3): 121–123. doi:10.1109/75.661135

Juhel S (2000) RF power transistors comparative study of LDMOS versus bipolar technology. ST microelectronics, Application Note AN1223

Karoui W, Parra T (2008) A protection circuit for HBT RF power amplifier under load mismatch conditions. Proc Int Northeast Workshop Circuits and Syst Conf, Montreal, Canada. pp 241–244. doi:10.1109/NEWCAS.2008.4606366

Kelly DE, Mekechuk K, Miller T (2008) Switch-mode power amplifier linearization. Proc Radio Freq Integr Circuits Symp, Atlanta, USA. pp 153–156. doi:10.1109/RFIC.2008.4561407

Ketola J, Sommarek J, Vankka J, Halonen K (2004) Transmitter utilizing bandpass delta-sigma modulator and switching mode power amplifier. Proc Int Symp Circuits Syst, Vancouver, Canada. pp 633–637. doi:10.1109/ISCAS.2004.1328274

Kitchen J, Chu WY, Deligoz I, Kiaei S, Bakkaloglu B (2007) Combined linear and modulated switched-mode PA supply modulator for polar transmitters. Int Solid-State Circuits Conf Dig Tech Pap, San Francisco, USA. pp 82–83. doi:10.1109/ISSCC.2007.373598

Knaipp M, Park JM, Vescoli V, Roehrer G, Minixhofer R (2006) Investigations on an isolated lateral high-voltage n-channel LDMOS transistor with a typical breakdown of 150 V. Proc Int Solid-State Circuits Conf, Montreux, Switzerland. pp 266–269. doi:10.1109/ESSDER.2006.307689

Koeppe J, Harjani R (2004) Enhanced analytic noise model for RF CMOS design. Proc Cust Integr Circuits Conf, Orlando, USA. pp 383–386

Marbell MN, Cherepko SV, Madjar A, Hwang JCM, Frei M, Shibib MA (2004) An improved large-signal model for harmonic-balance simulation of Si LD-MOSFETs. Proc Eur Microw Conf, Amsterdam, The Netherlands. pp 225–228

McCune E (2005) Polar modulation and bipolar RF power devices. Proc Bipolar/BiCMOS Technol Meet, Santa Barbara, USA. pp 1–5. doi:10.1109/BIPOL.2005.1555188

Metzger AG, Ramanathan R, Jiang L, Hsiang-Chih S, Cismaru C, Hongxiao S, Rushing L, Weller KP, Ce-Jun Wei, Yu Zhu, Klimashov A, Tkachenko YA, Bin L, Zampardi PJ, (2007) An InGaP/GaAs merged HBT-FET (BiFET) technology and applications to the design of handset power amplifiers. J Solid-State Circuits 42(10):2137–2148. doi:10.1109/JSSC.2007.904318

Milosevic D, van der Tang J, van Roermund A (2003) On the feasibility of applications of class E RF power amplifiers in UMTS. Proc Int Symp Circuits Syst, Bangkok, Thailand. vol 1, pp 149–152

MOBILIS FP6 IST Project (2006) Mixed SiP and SoC integration of power BAW filters for digital wireless transmissions. IST specific targeted research or innovation project No 027003, http://www.ist-mobilis.org

Mohapatra NR, Ehwald KE, Barth R, Rucker H, Bolze D, Schley P, Schmidt D, Wulf HE (2005) The impact of channel engineering on the performance and reliability of LDMOS transistors. Proc Eur Solid-State Device Res Conf, Grenoble, France. pp 481–484. doi:10.1109/ESSDER.2005.1546689

Muller D, Mourier J, Perrotin A, Szelag B, Monroy A (2005) Comparison of two types of lateral DMOSFET optimized for RF power applications. Proc Eur Solid-State Device Res Conf, Grenoble, France. pp 125–128. doi:10.1109/ESSDER.2005.1546601

Myung HG, Lim J, Goodman DJ (2006) Single carrier FDMA for uplink wireless transmission. Vehi Tech Mag 1(3):30–38. doi:10.1109/MVT.2006.307304

Pallotta A (2006) Quad-band GSM power amplifier by optimized BCD RFLDMOS. Top Workshop on Power Amplif Wirel Commun, San Diego, USA. pp 15–17

Pinon V, Hasbani F, Giry A, Pache D, Garnier C (2008) A single-chip WCDMA envelope reconstruction LDMOS PA with 130 MHz switched-mode power supply. Proc Int Solid-State Circuits Conf, San Francisco, USA. pp 564–636. doi:10.1109/ISSCC.2008.4523308

Presti CD, Carrara F, Palmisano G (2008) A high-resolution 24-dBm digitally-controlled CMOS PA for multi-standard RF polar transmitters. Proc Eur Solid-State Circuits Conf, Edimburgh, UK. pp 482–485. doi:10.1109/ESSCIRC.2008.4681897

Pritiskutch J, Hanson B (2000) Related LDMOS device parameters to RF performance. ST microelectronics, Application Note AN1228

Raab FH (1987) Efficiency of Doherty RF power-amplifier systems. Trans Broadc, BC-33:77–83. doi:10.1109/TBC.1987.266625

Ralph S, Farrell R (2007) Using high pass sigma-delta modulation for class-S power amplifiers. Proc Eur Conf Circuit Theory Des, Seville, Spain. pp 707–710. doi:10.1109/ECCTD.2007.4529694

Reynaert P, Steyaert MSJ (2005) A 1.75 GHz polar modulated CMOS RF power amplifier for GSM-EDGE. J Solid-State Circuits 40(12):2598–2608. doi:10.1109/JSSC.2005.857425

van Rijs F, Theeuwen SJCH (2006) Efficiency improvement of LDMOS transistors for base stations: towards the theoretical limit. Proc Electron Devices Meet, San Francisco, USA. pp 1–4. doi:10.1109/IEDM.2006.346998

Rinaldi N, d'Alessandro V, De Paola FM (2006) Electrothermal phenomena in bipolar transistors and ICs: analysis, modelling, and simulation. Proc Bipolar/BiCMOS Technol Meet, Maastricht, The Netherlands. pp 1–8. doi:10.1109/BIPOL.2006.311153

Scuderi A, Scuderi A, Carrara F, Palmisano G (2004) VSWR-protected silicon bipolar power amplifier with smooth power control slope. Proc Int Solid-State Circuits Conf, San Francisco, USA. vol 1, pp 194–522. doi:10.1109/ISSCC.2004.1332660

Shi B, Sundström L (2000) A 200-MHz IF BiCMOS signal component separator for linear LINC transmitters. J Solid-State Circuits 35(7):987–993. doi:10.1109/4.848207

Shimizu T, Matsunaga Y, Sakurai S, Yoshida I, Hotta M (2005) A single-chip Si-LDMOS power amplifier for GSM. Proc Int Solid-State Circuits Conf, San Francisco, USA. pp 310–312. doi:10.1109/ISSCC.2005.1493993

Sinnesbichler FX, Olbrich GR (1999) Electro-thermal large-signal modelling of SiGe HBTs. Proc Eur Microw Conf, Munich, Germany. pp 125–128. doi:10.1109/EUMA.1999.338426

Sirois J, Boumaiza S, Helaoui M, Brassard G, Ghannouchi FM (2005) A robust modelling and design approach for dynamically loaded and digitally linearized Doherty amplifiers. Trans Microw Theory Tech 53(9):2875–2883. doi:10.1109/TMTT.2005.854257

Skotnicki T, Bœuf F (2003) Introduction à la Physique du transistor MOS. In: Gautier J (ed) Physique des dispositifs pour circuits intégrés silicium. Lavoisier, Paris

Sowlati T, Rozenblit D, Pullela R, Damgaard M, McCarthy E, Koh D, Ripley D, Balteanu F, Gheorghe I (2004) Quad-band GSM/GPRS/EDGE polar loop transmitter. J Solid-State Circuits 39(12):2179–2189. doi:10.1109/JSSC.2004.836335

Spirito M, van der Heijden MP, Pelk M, de Vreede LCN, Zampardi PJ, Larson LE, Burghartz JN (2005) Experimental procedure to optimize out-of-band terminations for highly linear and power efficient bipolar class-AB RF amplifiers. Proc Bipolar/BiCMOS Technol Meet, Santa Barbara, USA. pp 112–115. doi:10.1109/BIPOL.2005.1555212

Srirattana N, Raghavan A, Heo D, Allen PE, Laskar J (2005) Analysis and design of a high-efficiency multistage doherty power amplifier for wireless communications. Trans Microw Theory Tech 53(3, Part 1):852–860. doi:10.1109/TMTT.2004.842505

Su DK, McFarland WJ (1998) An IC for linearizing RF power amplifiers using envelope elimination and restoration. J Solid-State Circuits 33(12):2252–2258. doi:10.1109/4.735710

Szelag B, Muller D, Mourier J, Arnaud C, Bilgen H, Judong F, Giry A, Pache D, Monroy A (2006) High RF performances asymmetric spacer NLDMOS integration in a 0.25 μm SiGe:C BiCMOS Technology. Proc Bipolar/BiCMOS Technol Meet, Maastricht, The Netherlands. pp 1–4. doi:10.1109/BIPOL.2006.311114

Theeuwen SJCH, Sneijers WJAM, Klappe JGE, de Boet JAM (2008) High voltage RF LDMOS technology for broadcast applications. Proc Eur Microw Integr Circuits Conf, Amsterdam, The Netherlands. pp 24–27. doi:10.1109/EMICC.2008.4772219

Tseng PD, Zhang L, Gao GB, Chang MF (2000) A 3-V monolithic SiGe HBT power amplifier for dual-mode (CDMA/AMPS) cellular handset applications. J Solid-State Circuits 35(9): 1338–1344. doi:10.1109/4.868045

Uang R, Keyzer J, Dalvi A, Sugiyama Y, Iwamoto M, Galton I, Asbeck P (2002) RF pulse modulation and the digitally driven class C power amplifier. Top Workshop on Power Amplif Wirel Commun, La Jolla, USA

Walling J, Lakdawala H, Palaskas Y, Ravi A, Degani O, Soumyanath K, Allstot D (2008) A 28.6 dBm 65 nm class-E PA with envelope restoration by pulse-width and pulse-position modulation. Proc Int Solid-State Circuits Conf, San Francisco, USA. pp 566–636. doi:10.1109/ISSCC.2008.452330

Wang F, Rutledge DB (2004) 60-W L-band class-E/Fodd,2 LDMOS power amplifier using compact multilayered baluns. Top Workshop Power Amplif Wirel Commun, San Diego, USA

Wang F, Kimball D, Popp J, Yang A, Lie DYC, Asbeck P, Larson L (2005) Wideband envelope elimination and restoration power amplifier with high efficiency wideband envelope amplifier for WLAN 802.11 g applications. Microw Theory Tech Symp Dig, Long Beach, USA. doi:10.1109/MWSYM.2005.1516688

Wang L, Wang J, Li R, Lee P, Hu J, Qu W, Li W, Yang S (2008) Novel STI scheme and layout design to suppress the kink effect in LDMOS transistors. Semicond Sci Technol 23(7):075025. doi: 10.1088/0268-1242/23/7/075025

Wang PC, Huang KY, Kuo YF, Huang MC, Lu CH, Chen TM, Chang CJ, Chan KU, Yeh TH, Wang WS, Lin YH, Lee CC (2008) A 2.4 GHz +25 dBm P-1dB linear power amplifier with dynamic bias control in a 65-nm CMOS process. Proc Eur Solid-State Circuits Conf, Edimburgh, UK. pp 490–493. doi:10.1109/ESSCIRC.2008.4681899

Yamamoto K, Suzuki S, Mori K, Asada T, Okuda T, Inoue A, Miura T, Chomei K, Hattori R, Yamanouchi M, Shimura T (2000) A 3.2-V operation single-chip dual-band AlGaAs/GaAs HBT MMIC power amplifier with active feedback circuit technique. J Solid-State Circuits 35(8):1109–1120. doi:10.1109/4.859499

Yousefzadeh V, Alarcón E, Maksimović D (2005) Efficiency optimization in linear-assisted switching power converters for envelope tracking in RF power amplifiers. Proc Int Symp Circuits Syst, Kobe, Japan. 1302–1305. doi:10.1109/ISCAS.2005.1464834

Zampardi PJ (2008) GaAs technology status and perspectives for multi-band and multi-standard challenges in upcoming RF-frontends. Proc Radio Week Symp, Orlando, USA. pp 187–190. doi:10.1109/RWS.2008.4463460

van Zeijl PTM, Collados M (2007) A digital envelope modulator for a WLAN OFDM polar transmitter in 90-nm CMOS. J Solid-State Circuits 42:2204–2211. doi:10.1109/JSSC.2007.905239

Zhang X, Larson LE, Asbeck PM, Langridge RA (2002) Analysis of power recycling techniques for RF and microwave outphasing power amplifiers. Trans Circuits and Syst–II: Analog Digit Signal Process 49(5):312–320. doi:10.1109/TCSII.2002.801411

Zhu Y, Gerber J, Cai Q (2002) Simulating multi-finger power HBTS. Microw J, vol 45, 3:96–104

Chapter 2
Discretized Reconfiguration Techniques for Radiofrequency Power Amplifiers

Abstract This chapter will be dedicated to efficiency enhancement based on *PA* fragmentation techniques. Compared to other architectures, the strength of such principle lies in a very broad power range over which a power amplifier can be reconfigured. This range is determined by the level of granularity and complexity that is affordable for the targeted application. If dynamic reconfiguration is controlled at the envelope rate, significant efficiency improvement can be expected for high *PAPR* applications such as *WIFI*, *WIMAX* or *LTE*. However, two fundamental drawbacks may arise from this technique. First, if the transmitted signal presents severe phase discontinuities, the receiver may encounter difficulties to properly track synchronization whereby the radio link can be interrupted (see section "Description of European 3G: *WCDMA* and Its Uplink High Data-Rate Extension *HSUPA*" in Chapter 1, this volume). Second, quantization noise results in a spurious level regrowth. *Error Vector Magnitude* is degraded and/or tests on emission spectral masks no longer pass. The following discussion will try to demonstrate that these issues can be circumvented providing some control/compensation circuitry is integrated with the power core itself.

Keywords Fragmented power amplifiers · Power stage bypass/extinction · Quantization noise · Delta-Sigma analog to digital conversion · Built-In Current Sensing (*BICS*) · Power detection · Spectral masks · Closed-loop *EVM* control · Variable-conversion-gain mixer

2.1 Introduction on Fragmented Power Amplifiers

The power amplifier can be split into power stages according to a cascaded topology (Fig. 2.1) and/or a parallelized topology (Fig. 2.2).

The reconfiguration power ranges of a dual cascaded/parallelized fragmented *PA* are illustrated in Fig. 2.3 for various operation modes.

When active, each power stage operates in its *optimum* regime, either in class E (Presti et al. 2008) or class AB (Deng et al. 2005). In the former case, efficiency is privileged but envelope restoration is necessary. Moreover, the response P_{OUT} vs.

L. Leyssenne et al., *Reconfigurable RF Power Amplifiers on Silicon for Wireless Handsets*, Analog Circuits and Signal Processing, DOI 10.1007/978-94-007-0425-1_2,

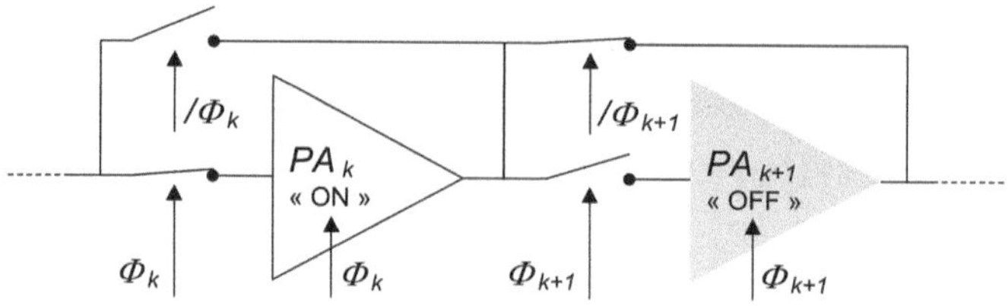

Fig. 2.1 *PA* architecture based on cascaded (bypassed) fragmented topology

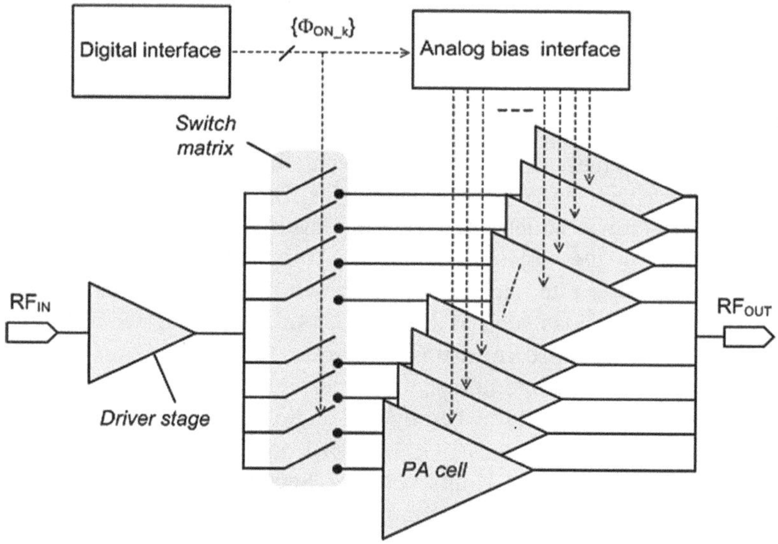

Fig. 2.2 *PA* architecture based on parallelized fragmented topology

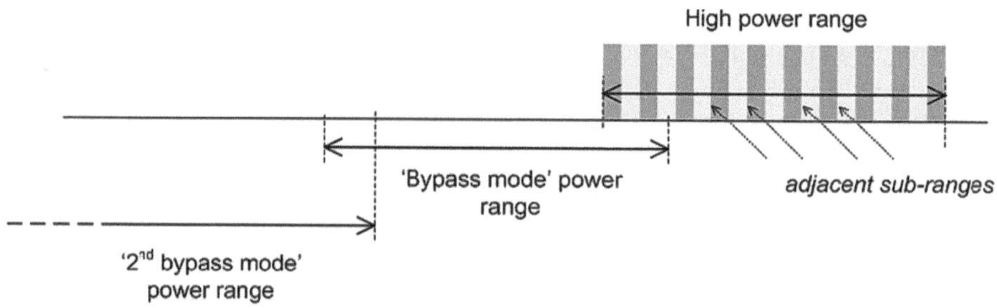

Fig. 2.3 Power ranges addressed by fragmented *PA*

P_{IN} is roughly stair-shaped which results in relatively high quantization noise. That is why the latter case is considered throughout Chapter 2. Indeed, side effects (at reconfiguration range borders) are therefore reduced and the response P_{OUT} vs. P_{IN} does not present too steep discontinuities.

Based on the software *Advanced Design System*, a system-level design methodology will be dissected into four different sections. Section 2.2 will present a slow reconfiguration-rate bypass technique that prevents strong phase steps. Sections 2.3, 2.4 and 2.5 will investigate fast reconfiguration-rate parallelized architectures whose goal is to track envelope variations in an over-sampled mode. A system-level design methodology will be proposed based on *Advanced Design System*. Over-sampling mitigates the benefits of *Envelope* analysis since a huge amount of data and consumption time are required for a single frame completely simulated. Therefore, using generic behavioral models (e.g. Saleh model for a *PA*) may be interesting. However, this may result in significant inaccuracies in linearity and efficiency. The behavioral modelling of secondary blocks (comparators) is affordable. All critical blocks (Power core, actuators, detectors...) should be designed at transistor level. In order to save design time, a good understanding of the issues and trade-offs underlying in such reconfigurable *PA* architecture is necessary. Figure 2.4 sums up the cross-interactions between the architecture parameters. Linearity requirements determine the fundamental architecture parameters, i.e. over-sampling ratio and resolution. The higher these factors are, the better the linearity will be. The dynamic power range over which reconfiguration is targeted also rules the performances of the employed envelope detector and the resolution M as shown by (2.1).

$$M \geq \frac{\ln\left(\frac{P_{\max}}{P_{\min}}\right)}{\ln(2)} = \left(\frac{P_{\max\_dB} - P_{\min\_dB}}{10}\right) \times \frac{\ln(10)}{\ln(2)} \tag{2.1}$$

Suppose that *PAPR* is a relevant evaluation of the power ratio, the necessary resolution for a *WLAN* application is $M = PAPR \times \ln(10)/10/\ln(2) = 4$. In order to cancel out noise over the operation bandwidth (namely 80 MHz for *WIFI*), the cut-off

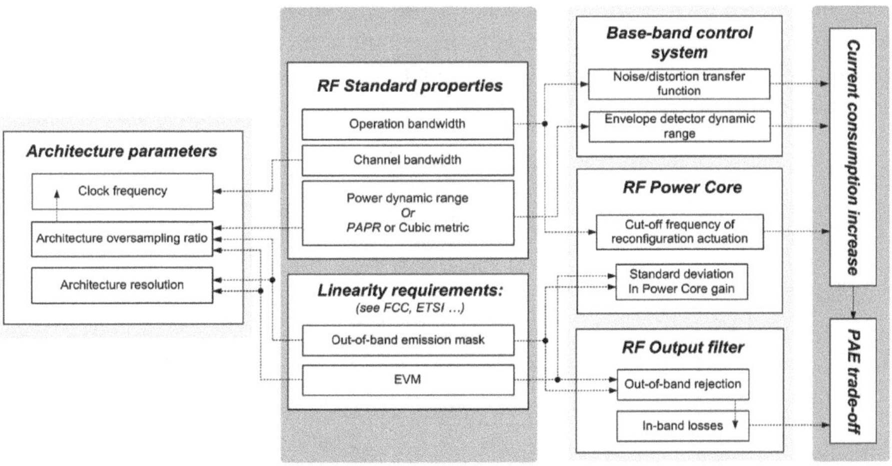

Fig. 2.4 Illustration of design methodology: issues/trade-offs

frequency of the base-band control systems must be dictated accordingly. The cut-off frequency of the power core reconfiguration actuators must be set high enough to have marginal effect on the overall architecture response time (>10 × 80 MHz for *WIFI*). Linearity performance will be all the higher as the power gain deviation is minimal (in magnitude and phase) over the whole reconfiguration range. The output filter rejection in the vicinity of the operation bandwidth also positively impacts linearity. In an opposite way, the need for high efficiency imposes a reduction of the output filter in-band losses (resp. current consumption) whereby the affordable out-of-band rejection (resp. the affordable bandwidth of analogue blocks) is limited. The output filter response is determined by the employed duplex method as well. At first sight, the use of fragmented power amplifiers is more recommended for applications based on *TDD* rather than *FDD* duplex method (such as e.g. *WCDMA*) unless a sufficiently selective duplexer output filter is used. From the point of view of *TX* efficiency, the development of *BAW* resonators is a promising technological enhancement since they present high quality factors and withstand high power densities whereby the complexity of fragmented power amplifiers (resolution, *OSR*, reconfiguration rate. . .) is loosened.

2.2 Power Amplifier Bypass Technique

2.2.1 Introduction

The *PA* topology presented here has been proposed in the frame of the *RNRT* project *Asturies*, under a national grant from the French Research Ministry. The aim of this project was to develop *RF* and base-band architectures for reconfigurable 3rd Generation terminals, *DCS* and *WCDMA*. The linearity/efficiency trade-off is dealt by means of a power stage that is bypassed and switched off at low power levels to increase battery lifetime. The general bypass-based *PA* synoptic is given in Fig. 2.5. The hereunder architecture is based on the association of a differential power amplifier and a differential driver. This driver combines several functionalities.

On the one hand, it plays the role of a cascode amplifier that ensures unilateral response for the overall *PA*. On the other hand, it behaves as an *Emitter Coupled*

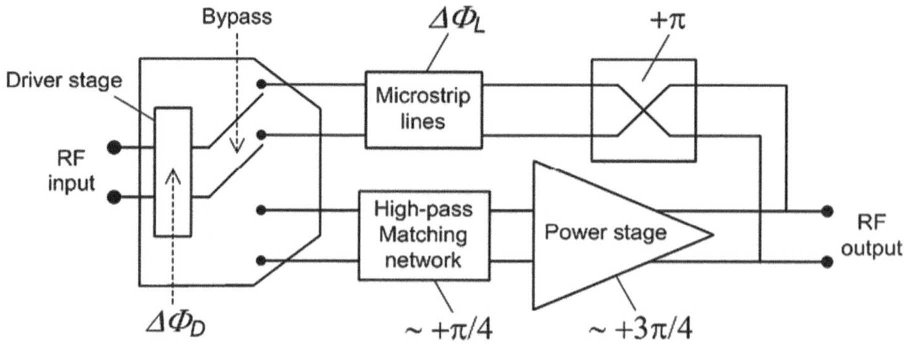

Fig. 2.5 Bypassed-based *PA* synoptic

Logic-based switch. An enable control voltage En_{bypass} allows transferring the *RF* signal to the power stage input when *OFF*, or directly to the power stage output when *ON*. Moreover, the power stage bypass is coupled with the extinction of the power stage in order to increase battery life-time. The transmitter *AGC* loop (not depicted here) is aimed to compensate for the sudden *PA* power gain collapse (approximately 20 dB) and quickly increase the power level at the *PA* input. Nevertheless, as it was explained in Section 1.1.3.1, the *AGC* loop only applies on output magnitude and phase steps are not corrected. That is the reason why the *PA* must fulfill the phase hopping requirement on its own (Tables 1.2 in Chapter 1) so as to increase the maximum allowed bypass rate.

2.2.2 Bypass Topology

Figure 2.5 summarizes the forward transfer phase budget over the various elements. The inter-stage matching network between the driver and power stages is a high-pass network, i.e. a shunt inductor in association with a series capacitor. At resonance, it features a $\pi/4$ transmission phase shift. Additionally, the phase of a matched power stage is practically in the range $3\pi/4\pm\Delta\Phi_P$ according to the input quality factor. In this condition, the transmission phase of the high power path (i.e. the cascaded inter-stage matching network and power stage) is in the vicinity of π. To cope with the previously mentioned phase step specification, differential topology is found to be of great profit. Indeed, by cross-coupling positive and negative nodes in the *bypassed* path, it becomes possible to introduce a π phase shift that roughly matches the phase of the power stage transfer function. An additional controllable phase correction is necessary to guarantee that the reconfiguration phase step might be confined in the specified range [0–30°] for any *VSWR* condition. To this end, a switched feedback network (R_1, C_1, L_2) is included in the driver stage. The use of L_2 is to increase the feedback phase margin and therefore stability at high frequency. The phase correction due to feedback is as high as 10°. This is at the cost of a 5 dB power gain decrease.

The detailed schematic of the *PA* architecture is depicted in Fig. 2.6 (Leyssenne et al. 2005b). It is based on high voltage *HBT* transistors with base and emitter degeneration in compliance with the calculation and discussion of Chapter 1. A rule of thumb in order to avoid *Kirk effect* must consist in biasing all transistors 10% below the current density that gives maximum transition frequency.

The power stage is a quasi-common emitter differential stage and consumes a 500 mA quiescent current. The power transistors $Q_{4\pm}$ feature 250×5 μm emitter lengths. The inter-stage matching network is made with the association $\{L_3 = 3.5\ \text{nH}; C_3 = 10\ \text{pF}\}$ in order to match the 200 Ω driver output impedance to the 20 Ω power stage input impedance. For sake of clarity, the inter-stage external DC-feed inductors are not depicted in Fig. 2.6. The driver stage is a cascode amplifier and consumes a 50 mA quiescent current. The common base transistors $Q_{2\pm}$ and $Q_{3\pm}$ play the role of switches and feature a 25 × 5 μm emitter length. The driver input matching is eased by the inductive degeneration (L_1) of the transistors $Q_{1\pm}$. (25 × 5 μm emitter length) The real part of the input impedance is given by (2.2):

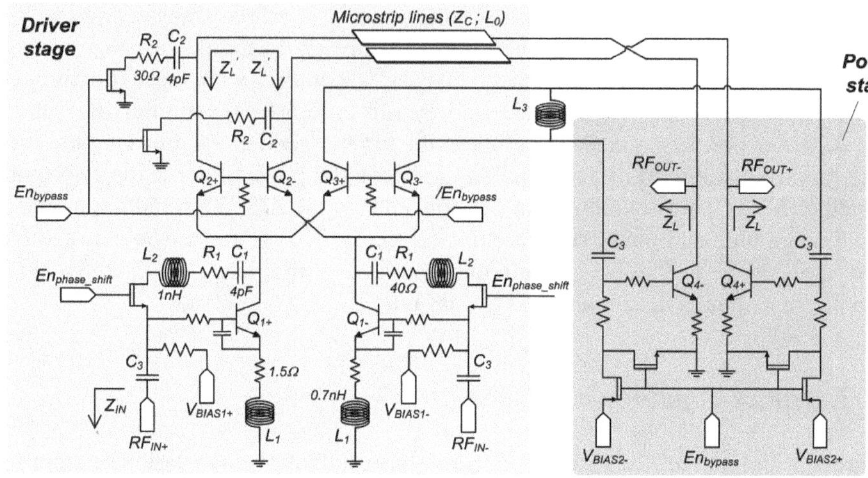

Fig. 2.6 Bypassed-based *PA* schematic

$$\Re\left(Z_{IN}\right) \approx \omega_T \cdot L_1 \tag{2.2}$$

In the frame of *Asturies* project, for compliance purpose with the *I/Q* modulator output impedance, the targeted Z_{IN} value is 100 Ω. The resulting L_1 inductance value is therefore 0.7 nH. One of the issues that this topology has to address is to ensure a correct output impedance matching both in full-power and in bypass modes. For a linear class A differential amplifier, the output optimum impedance is determined by the required output power (27 dBm):

$$Z_L = \frac{\left(V_{CC} - V_{headroom}\right)^2}{P_{OUT}} \tag{2.3}$$

where V_{CC} is the supply voltage, and $V_{headroom}$ is the voltage margin on the output swing in order to prevent the power transistor from saturating.

Assuming 27 dBm output power is targeted with a 3.5 V power supply and a 0.9 V headroom voltage, the output load Z_L value is 12 Ω. When the *bypass* mode is selected, the output impedance of the driver stage is in the order of 100 Ω. Therefore, the output impedance mismatch in bypass mode must be reduced, under the constraint that it should leave the output matching unchanged when the full power mode is active. To this end, the lossy shunt network (R_2,C_2) is switched on in *bypass* mode.

Due to the relatively large power stage dimensions (750 × 750 μm^2), the impact of the *bypass* output microstrip may not be neglected in high power mode. The resulting impedance Z_L is given by (2.4):

$$Z_L = Z_0 \times \frac{Z'_L\big|_{bypass\_OFF} + j Z_0 \tan\left(\frac{2\pi\, l_0}{\lambda}\right)}{Z_0 + j\, Z'_L\big|_{bypass\_OFF} \cdot \tan\left(\frac{2\pi\, l_0}{\lambda}\right)} \approx -j \cdot \frac{Z_0\, \lambda}{2\pi\, l_0} \tag{2.4}$$

where Z'_L is the driver output impedance at *bypass* access (when bypass is *OFF*), Z_0 and l_0 are respectively the microstrip characteristic impedance and length.

In order to prevent electromigration, the microstrip width must not be chosen below 20 μm. The characteristic impedance is then as high as 50 Ω. The wavelength λ at 2 GHz on silicon is 75 mm. The microstrip lines introduce an extra phase $\Delta\Phi_L$ that does not exceed 4° and an equivalent capacitance that is $C_{eq} = l_0/(\lambda \times f \times Z_0) = 100$ fF at *PA* output.

Figure 2.7 depicts the simulation results in terms of S_{21} with the presented topology.

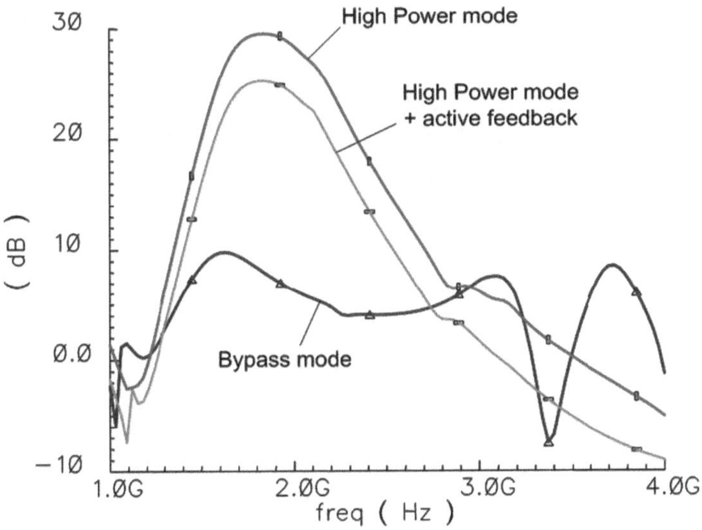

Fig. 2.7 Power amplifier S_{21}

Figure 2.8 depicts the large signal performances of this topology in various modes.

The compression point P_{-1dB} is 27 dBm (resp. 12 dBm) in high power mode (resp. *bypass* mode). The maximum *PAE* is as high as 55% at $P_{OUT} = 29$ dBm in high power mode (resp. *bypass* mode). At medium power level (at $P_{OUT} = 11$ dBm), switching to bypass mode allows enhancing *PAE* from 1 to 15% which demonstrates the benefit of this topology.

2.2.3 Experimental Results

A *PA bypass* demonstrator was realized with ST Microelectronics 0.25 μm *SiGe* technology. The *PA* demonstrator was enclosed in a *TQFP* package and connected on an Epoxy board. For test purposes, external input/output baluns were employed. *LC* topology was employed at the input port (Fig. 2.9) whereas ring topology was chosen at the output port (Fig. 2.10b) for wider bandwidth purpose. *LC* balun elements are designed according to the (2.5), i.e. $L_{balun} = 8.2$ nH, $C_{balun} = 0.82$ pF.

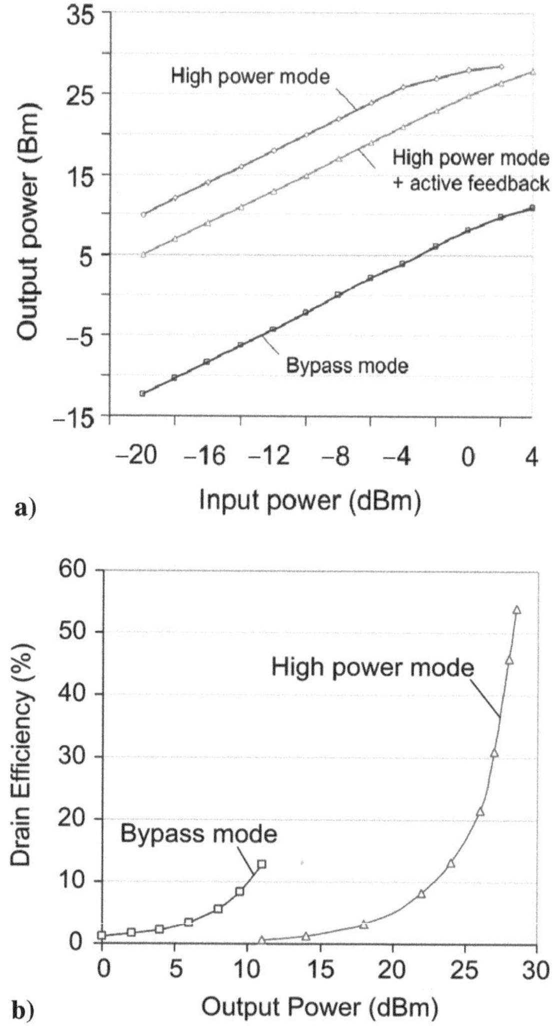

Fig. 2.8 (**a**) Output power vs. input power, (**b**) power added efficiency vs. input power in various modes (high power, high power + active feedback, bypass modes)

$$
\begin{aligned}
L_{balun} &= \frac{\sqrt{R_{diff} \times R_{single}}}{2\pi f} \\
C_{balun} &= \frac{1}{2\pi f \times \sqrt{R_{diff} \times R_{single}}}
\end{aligned} \tag{2.5}
$$

Figure 2.10a, b respectively depict a die micrograph of the wire-bonded and packaged Power amplifier (in a slugged down *TQFP* package) and a test *PCB* photograph. The *S*-parameters measurement are depicted in Figs. 2.11 and 2.12.

The *PA* presents a 20.3 dB power gain in high power mode and a 0.4 dB in bypass mode. Power gain losses up to 4 dB both in high power and bypass modes

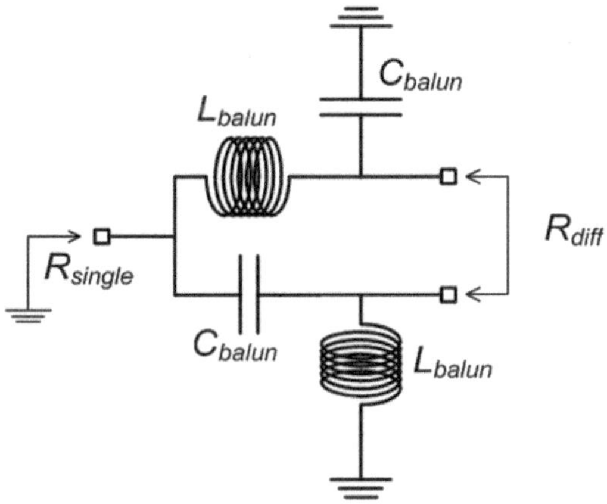

Fig. 2.9 LC balun topology

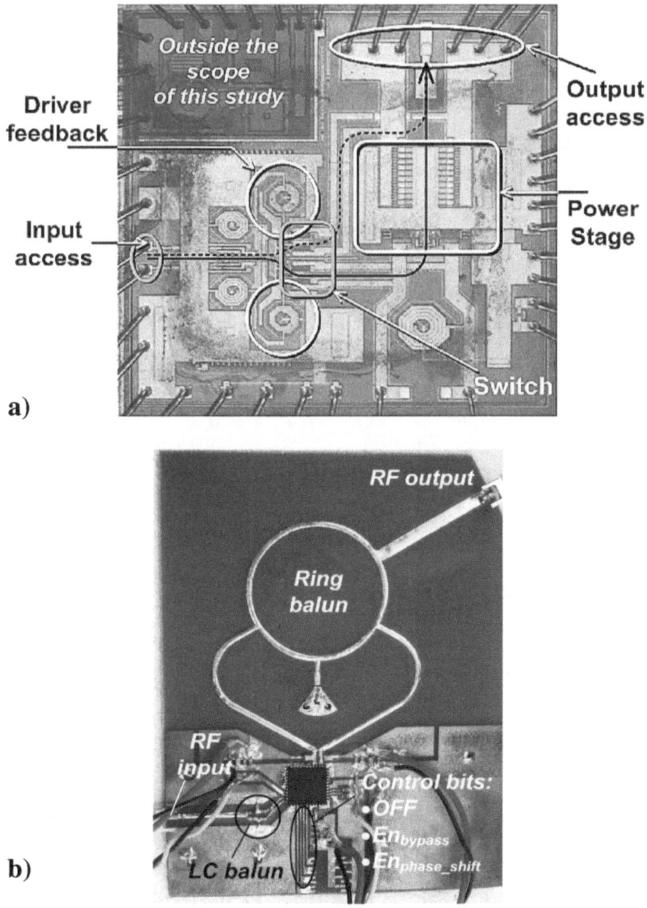

Fig. 2.10 (**a**) Bypass *PA* die micrograph (2.31 × 2.38 mm^2), (**b**) test board photograph

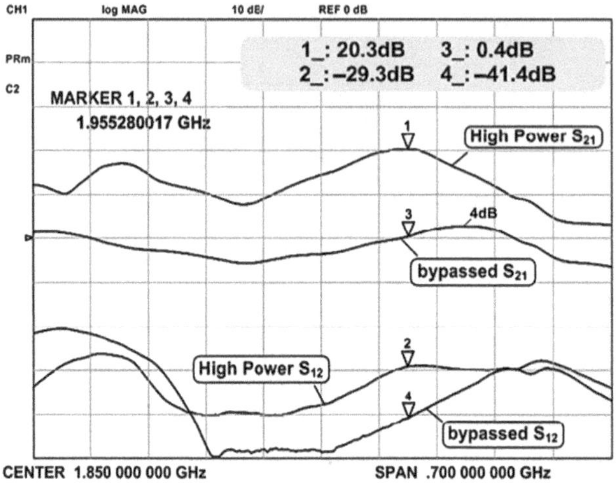

a) S_{21}, S_{12}

Fig. 2.11 Bypass *PA* transfer gain S_{21} and isolation S_{12}

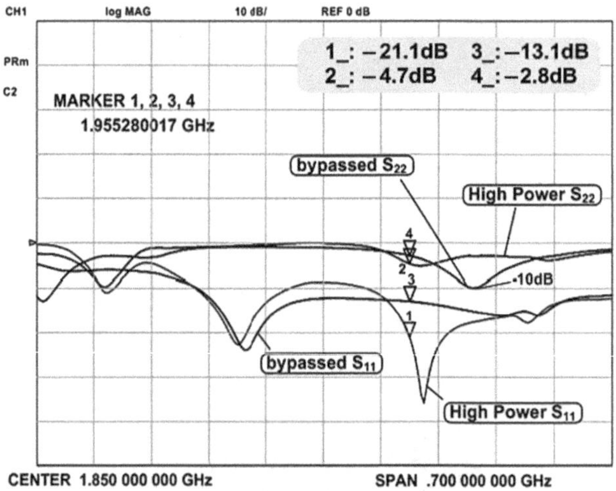

b) S_{11}, S_{22}

Fig. 2.12 Bypass *PA* return losses S_{11} ad S_{22}

are observed due to a frequency shift in the output balun response that increases its unbalance and degrades S_{22} return loss to –5 dB (resp. –3 dB) in high power mode (resp. in bypass mode). The input *LC* balun results in a –21 dB (resp. –13 dB) S_{11} return loss in high power mode (resp. bypass mode).

The large-signal response of the dual mode power amplifier is depicted in Fig. 2.13. The *PA* respectively features a 23.3 dBm and a 6.6 dBm compression points in high power and bypass modes (i.e. 4 dB below the simulated level in both modes, probably for the same unbalance reason that was previously put forward

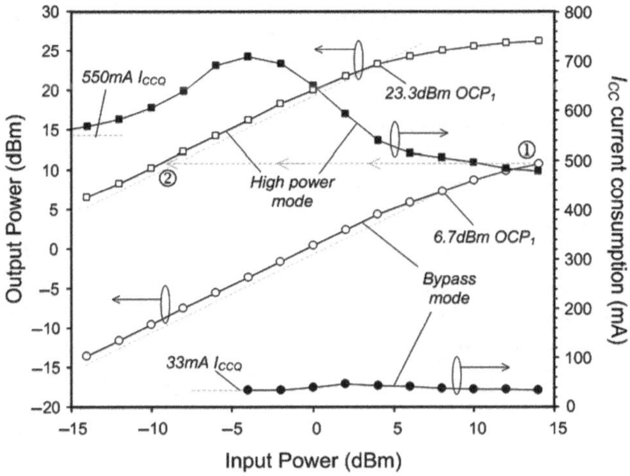

Fig. 2.13 Measured output power and average consumed current vs. input power

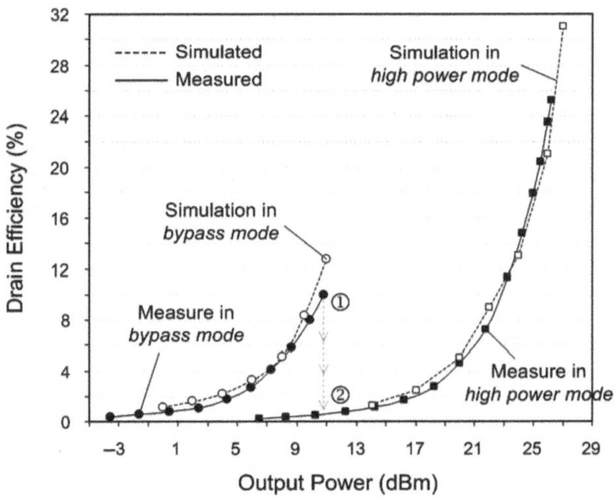

Fig. 2.14 Measured output collector efficiency vs. output power

about power gain). The dotted lines in Fig. 2.13 illustrate the 22 dB gain step from point ① to point ② that occurs when the *PA* switches from one mode to the other (and which is compensated via *AGC* loop by a 22 dB step in input power).

According to Fig. 2.14, good agreement is observed between measures and simulations, and the *PA* respectively features a 25 and a 10% maximum *PAE* in high power and bypass modes as depicted in.

In analogy to Figs. 2.13 and 2.14 illustrates the 9% collector efficiency step from point ① to point ② due to the *PA* reconfiguration that occurs when the *PA* switches from one mode to the other.

2.3 Reconfigurable Power Amplifier Based on Parallelized Switched Power Cells

2.3.1 Introduction on Discretized Power Amplifiers

A M-bit discretized power amplifier (Fig. 2.15) is built with $N = 2^M$ parallel power cells that can be alternately switched on/off by means of fast buffers. This topology was proposed in the frame of the European project *Uppermost–MEDEA+ 2A202* (Leyssenne et al. 2005a). Series switches are also inserted to provide sufficient isolation and ensure the immunity of switched-off cells on the *RF* behavior. The activation buffers must be designed so as to drive the low-impedance input of power cells with a response time that is compliant with the targeted application. Switches are activated via the control signals $\Phi_{ON,j}$. The normalized digital control word is defined as:

$$w_{ctrl} = \frac{1}{2^M \cdot V_{DD}} \cdot \sum_{j=1}^{2^M} \Phi_{ON,j} \in [0; 1] \tag{2.6}$$

where M is the architecture resolution.

As the digital word w_{ctrl} dynamically controls the *PA* capability, it must track the *RF* envelope variations as closely as possible. Dynamic power gain variations result in envelope distortion and should be minimized, both in magnitude and phase. When they are *ON*, cells operate at a fixed bias current density (class-A). This way,

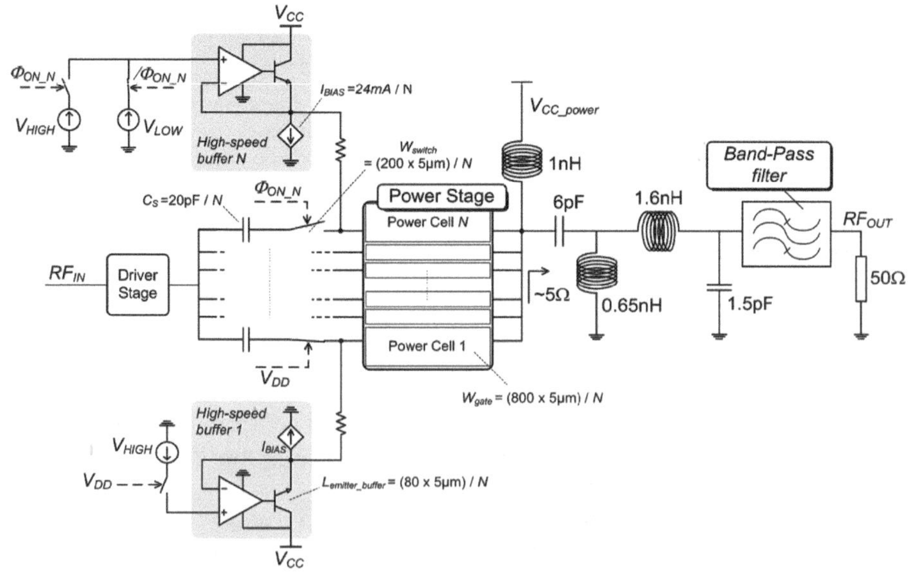

Fig. 2.15 Discretized power amplifier with parallel cells

as shown by (Deng et al. 2005), the power gain magnitude of the overall structure has little dependency over the number of active cells since it is mainly determined by transition frequency ω_T according to the following (2.7). This is confirmed by (B.7) and (B.8) in the mathematical development of Appendix B:

$$G_P(\omega) \approx 4\left|\frac{Z_L}{Z_S}\right| \times \left|\left(\frac{\omega_T}{\omega}\right) \times \frac{1}{\left(1+\omega_T\, Z_L\, C_{gd}\right)}\right|^2 \qquad (2.7)$$

where Z_L, Z_S are load and source impedances, C_{gd} is the drain/gate capacitance.

Meanwhile, input and output impedance matching conditions are not independent of the number of active cells. Therefore, output load and inter-stage matching should be designed so as to present minimized power gain variations among the various configurations while privileging good efficiency response whatever the used technology might be. As *WLAN* standard is targeted, the resonance frequency was chosen to be approximately 2 GHz. The inter-stage matching network is made of a 0.9 nH DC-feed shunt inductor (not depicted n Fig. 2.15) and an overall 20pF series capacitors C_S. In the case of a 3-bit power stage, the series capacitor of each power cell is 20 pF/2^3 = 2.5 pF. If the power cells are built with common-source *LDMOS* devices, the typical power gain standard deviation is in the order of 0.6 dB. Figure 2.16 depicts output power as a function of input power for various numbers of active *LDMOS* cells.

The dark line represents the interpolated power locus that results from the adaptive switching between the various power waves and features analogies with the response of an *ADC*. Indeed, it "follows" the full power response (when all cells are *ON*) except that power quantization steps are observed. The conversion of quantization noise to distortion is a key parasitic feature of such discretized architecture. Replacing class-A cells with class-E/F cells could be considered for efficiency purpose but would result in steeper power discontinuities, or in other

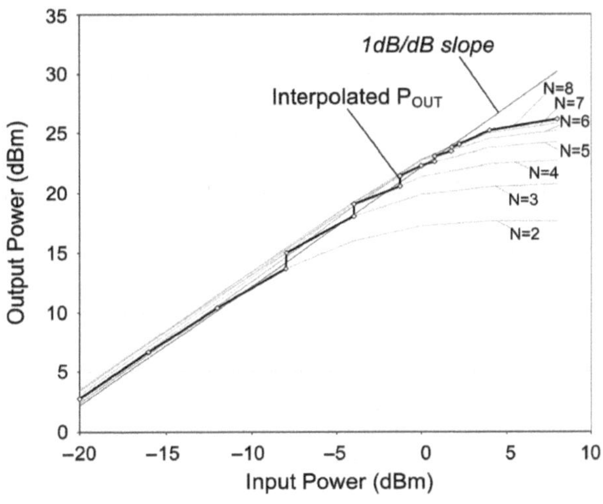

Fig. 2.16 Output power vs. input power for various number of active cells

words, in a decrease of the *Signal to quantization Noise/Distortion Ratio* (*SNDR*). Figure 2.16 was obtained with a *LDMOS* power stage whose overall gate width is 800×5 μm in order to achieve a maximum OCP_1 as high as 25 dBm. The actuation *HBT* buffers feature an overall emitter length of 80 × 5 μm and consume an overall 24 mA collector current to ensure that rise/fall times remain in the order of 1 ns (i.e. much shorter than a *WLAN* symbol duration). Series switches are *NMOS* devices whose overall gate width is 200 × 5 μm to reduce series resistive losses.

Identically, the output phase slightly depends on the number of active cells and its variations must be minimized. Figure 2.17 depicts the output phase of a *LDMOS* power stage as a function of the input power for various number of active cells.

The dark line represents the interpolated phase deviation locus that results from the adaptive switching between the various power waves. By proper design care, phase quantization steps are minimized though not completely removed. This phase hopping introduces an undesirable *phase modulation* f_{PM} whose effect will be developed further on:

$$f_{PM}(w_{ctrl}) = \exp\left(j\Delta\varphi\left(w_{ctrl}\right)\right) \tag{2.8}$$

From the point view of phase modulation, the use of *MOS* or *LDMOS* devices provides better performance than that of *HBT* transistors since the non-linearities in C_{gs} (resp. C_{gd}) are significantly lower than in C_π (resp. C_μ).

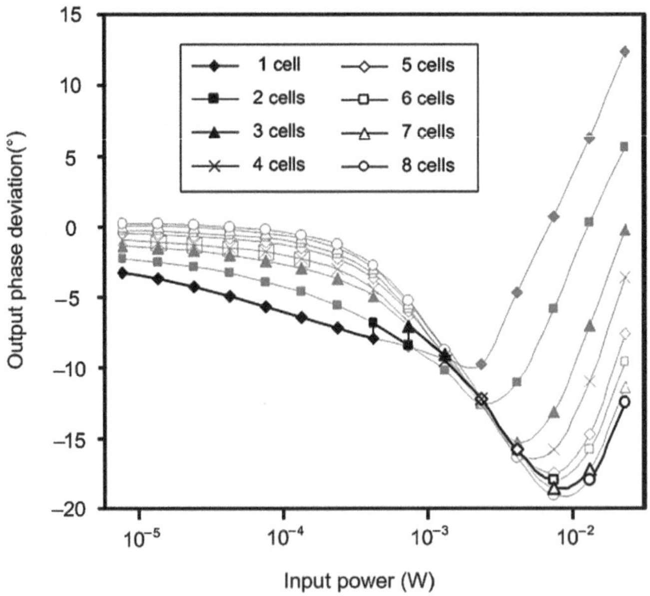

Fig. 2.17 Output phase deviation vs. input power for various numbers of active cells

2.3.2 Dynamic Modulation of Non-linear Kernels

Volterra formalism is generally considered as an accurate way of modelling complex non-linear circuits, especially those with memory effects (Cherry 1994, Zorn et al. 2008). Appendix C reviews the most fundamental theoretical equations that underlie Volterra formalism both in time and frequency domains. It is only valid for invariant systems, which is not the case of a dynamically reconfigurable power amplifiers. Therefore, in order to formalize the non-linear behavior of such *PA*, a few theoretical adaptations are needed. Practically, for sake of simplicity, jth-order non-linear kernels are considered as modulated by the digital control word in magnitude and phase according to the following interpolated expression:

$$g_{j,k} \approx g_{j,1} - \Delta g_j \cdot w_{ctrl} \cdot f_{PM}\left(w_{ctrl}\right) \tag{2.9}$$

where

$g_{j,1}$ is the jth order *Volterra* kernel when only one power cell is active,
$\Delta g_j = g_{j,1} - g_{j,2^M}$ is the maximum modulation of the *Volterra* kernel (between one active cell, and 2^M active cells),
$f_{PM}(w_{ctrl})$ is the phase modulation of the *Volterra* kernel, and
w_{ctrl} is the digital control word.

By substituting (2.9) in the classic *Volterra* series, the *PA* output signal in time domain can be expanded as follows:

$$\begin{aligned} v_{out}(t) = & \underbrace{\left(\sum_{j=1} \frac{1}{j!} \int_{-\infty}^{+\infty} du_1 \cdots \int_{-\infty}^{+\infty} du_j \times g_{j,1}\left(u_1, \cdots, u_j\right) \times \prod_{r=1}^{j} v_{in}\left(t - u_r\right)\right)}_{term\ A} \\ & \underbrace{- w_{ctrl}(t) \times f_{PM}(t) \times \left(\sum_{j=1} \frac{1}{j!} \cdot \int_{-\infty}^{+\infty} du_1 \cdots \int_{-\infty}^{+\infty} du_j \cdot \left(\Delta g_j\left(u_1, \cdots, u_j\right) \times \prod_{r=1}^{j} v_{in}\left(t - u_r\right)\right)\right)}_{term\ B} \end{aligned} \tag{2.10}$$

Similarly, the frequency domain *Volterra*-like series of the *PA* output signal is given by:

$$\begin{aligned} V_{OUT}(f) = & \underbrace{\left(\sum_{j=1}\left(\frac{1}{j!} \int_{-\infty}^{+\infty} df_1 \cdots \int_{-\infty}^{+\infty} df_j \cdot \left(\begin{matrix} G_{j,1}\left(f_1, \cdots, f_j\right) \times V_{IN}\left(f_1\right) \times \\ V_{IN}\left(f_2\right) \cdots V_{IN}\left(f - \sum_{r=1}^{j} f_r\right)\end{matrix}\right)\right)\right)}_{term\ A} \\ & \underbrace{- W_{CTRL}(f) \otimes F_{PM}(f) \otimes \left(\sum_{j=1} \frac{1}{j!} \int_{-\infty}^{+\infty} df_1 \cdots \int_{-\infty}^{+\infty} df_j \cdot \left(\begin{matrix}\Delta G_j\left(f_1, \cdots, f_j\right) \times V_{IN}\left(f_1\right) \times \\ V_{IN}\left(f_2\right) \cdots V_{IN}\left(f - \sum_{r=1}^{j} f_r\right)\end{matrix}\right)\right)}_{term\ B} \end{aligned} \tag{2.11}$$

where

$G_{j,k}$ is the jth order *Volterra* kernel in frequency domain with k active cells,
ΔG_j is the maximum deviation of the jth order *Volterra* kernel in frequency domain,
V_{IN} is the Fourier transform of the input signal spectrum,
W_{CTRL} is the Fourier transform of the digital control word w_{ctrl},
F_{PM} is the Fourier transform of the phase modulation function f_{PM}.

These *Volterra*-like series dedicated to dynamically adaptive circuits are composed of two components. The term A expresses the non-linearity of the *PA* if a single cell is only active (very low power mode) that is depicted by the IMD_3 level in Fig. 2.18a.

The second term B expresses the linearity enhancement, i.e. the IMD_3 reduction due to the dynamic *PA* reconfiguration as depicted in Fig. 2.18b, c. Meanwhile, such linearity improvement is mitigated by the quantization noise in the control word W_{CTRL} that is up-converted to the *RF* domain by convolution with the input *RF* signal. The control word w_{ctrl} is a digitalized image of the *RF* envelope and can be modeled by (2.12):

$$w_{ctrl}(t) = \alpha_{DET} \times v_{IN}^{\eta}(t) + q_n(t) \tag{2.12}$$

where q_n is the (white) quantization noise, and the factor α_{DET} and the exponent η allow modelling the envelope detection response.

According to the employed detector topology (logarithmic, *RMS*, and so on…), the factor η can present much variation. Sections 2.4.3.2 and 2.5.2.2 (resp. Sections 2.4.2.2 and 2.5.3.2) will detail some techniques that make η (resp. α_{DET}) adjustable.

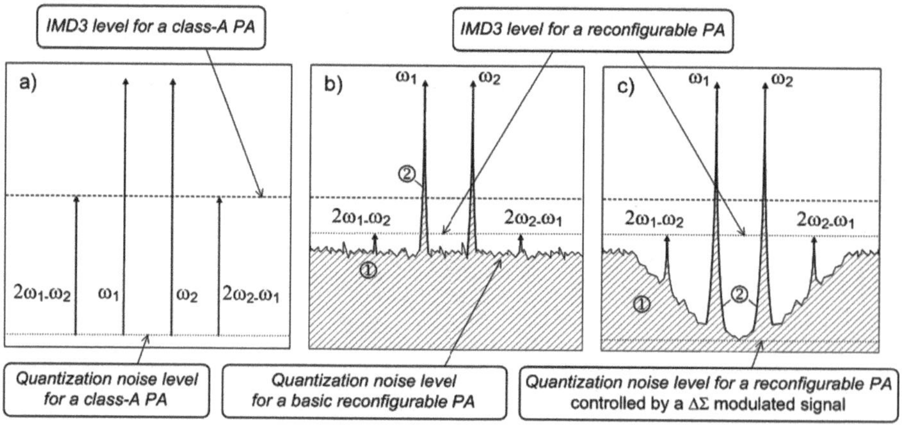

Fig. 2.18 Illustration of linearity enhancement via dynamic reconfiguration, and the non-linear drawbacks (①, and ②) due to quantization noise

If no attention is paid to noise drawback, severe white noise level regrowth can be encountered (see mark ① in Fig. 2.18b). Moreover, the phase modulation spectrum F_{PM} is a function of the noisy control word W_{CTRL} and introduces pink phase noise in the vicinity of carriers (see mark ② in Fig. 2.18b). Therefore, one of the fundamental goals of the architecture that will be presented in the following sections is to control and cancel the noise contribution in the operation bandwidth as depicted in Fig. 2.18c and increase the *SNDR*.

To make further theoretical developments, some assumptions must be made: 1st order gain deviation ΔG_1 is nil, the non-linear analysis is limited to 5th order, the exponent η is approximately 2. Phase modulation and memory effects are assumed to be neglected. In other words, $f_{PM} = 1$ and *Volterra* kernels are frequency independent:

$$\forall \left\{ \omega_1, \omega_2, \cdots, \omega_j \right\},\ \Delta G_j \left(\omega_1, \omega_2, \cdots, \omega_j\right) \approx \Delta G_j$$

Then by properly substituting (2.12) in (2.10), the *PA* output voltage can be expressed as a simple power series with regard to the input voltage as follows:

$$v_{out}(t) \approx g_{1,1} \times v_{in}(t) + g_{3,1} \times v_{in}^3(t) \times \left(\underbrace{1 + \left(\frac{g_{5,1} - \alpha_{DET} \Delta g_3}{g_{3,1}} \right) \times v_{in}^2(t)}_{C = 3^{rd}\ order\ mitigation\ function} - \frac{\alpha_{DET} \Delta g_3}{g_{3,1}} \times q_n \right) + o\left(|v_{in}(t)|^5 \right) \tag{2.13}$$

Third order nonlinearity can be reduced by minimizing the noise contribution q_n by means of *Delta-sigma* modulation techniques for example, and by mitigating the term into brackets C in (2.13) over the targeted power range. In frequency domain, the principle of discrete reconfiguration can be understood through means of Fig. 2.19.

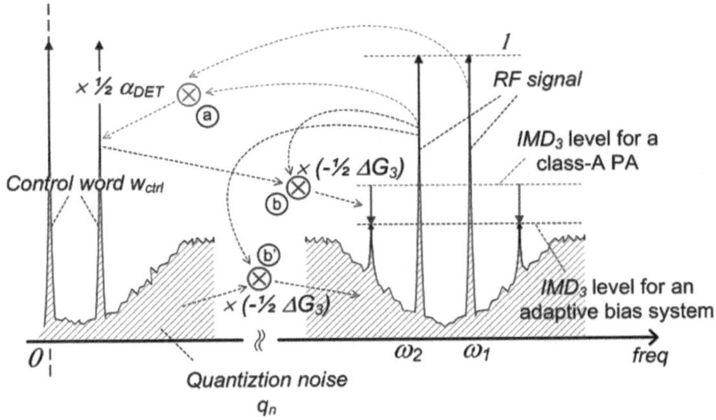

Fig. 2.19 Spectrum of *RF* signal $V_{OUT}(f)$, $W_{CTRL}(f)$

The envelope detector generates a 2nd order control word W_{CTRL} (mixing product (a)) that modulates the 3rd Volterra kernel and therefore convolutes with $V_{IN} \otimes V_{IN} \otimes V_{IN}$ (mixing product (b)). This behaves as if it has generated a 3th order harmonic and combated the 3rd degree non-linearity of the class-A *PA* (one active cell). Identically, quantization noise intervenes in the 3rd Volterra kernel modulation (mixing product (b′)) and degrades the purity of the *RF* spectrum.

2.4 Delta-Sigma Built-In Current Sensing in the Prospect of Power Amplifier Dynamic Reconfiguration

2.4.1 Delta-Sigma Modulation Basics

Many books and publications have been devoted to the principle and application of Delta-Sigma modulation. The basic principle consists in canceling the quantization noise introduced by analog to digital conversion in a bandwidth of interest. At this end, two techniques are combined. On the one hand, by over-sampling (relatively to *Nyquist*-rate sampling) the quantization white noise power is spread over a wider frequency range as depicted in Fig. 2.20.

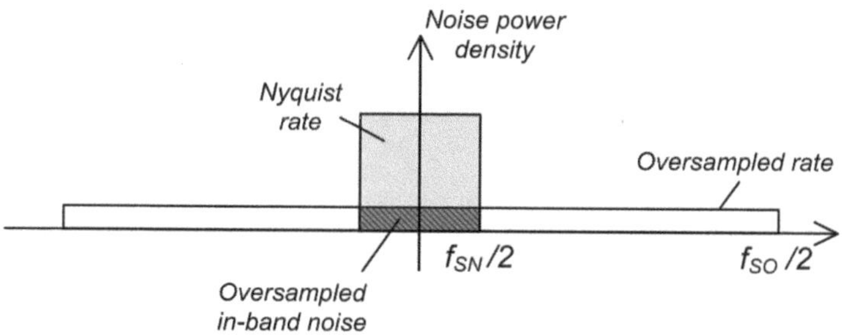

Fig. 2.20 Noise power density in Nyquist and oversampled cases (f_{sN} and f_{so} are respectively the Nyquist sampling frequency and the oversampling frequency)

On the other hand, the comparator output (where quantization noise appears) is fed back and differentiated with the input analog signal (Fig. 2.21). The resulting error signal is then "averaged" in classical topologies, which allows for the rejecting of the quantization noise that is in the useful bandwidth without increasing the comparator resolution (Fig. 2.22).

2.4.2 Power Detection via Delta-Sigma Built-In Current Sensing

2.4.2.1 Built-In-Current Sensor Principle

In modern cellular or *WLAN* applications, the reconfiguration of *RF* blocks according to traffic conditions is of great importance in order to dynamically loosen the

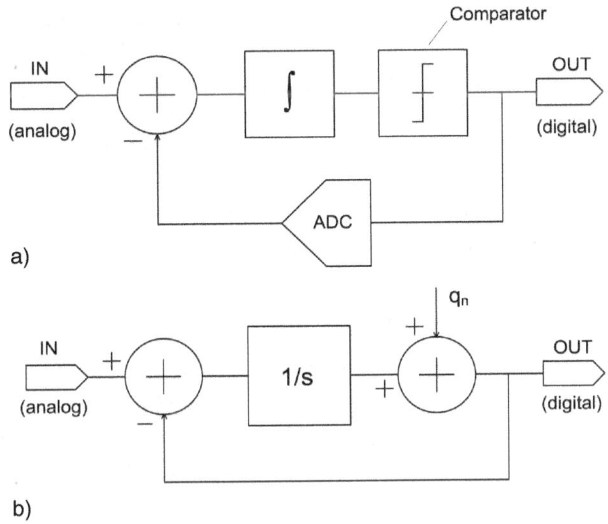

Fig. 2.21 (**a**) 1st order delta/sigma modulator, (**b**) 1st order delta/sigma synoptic

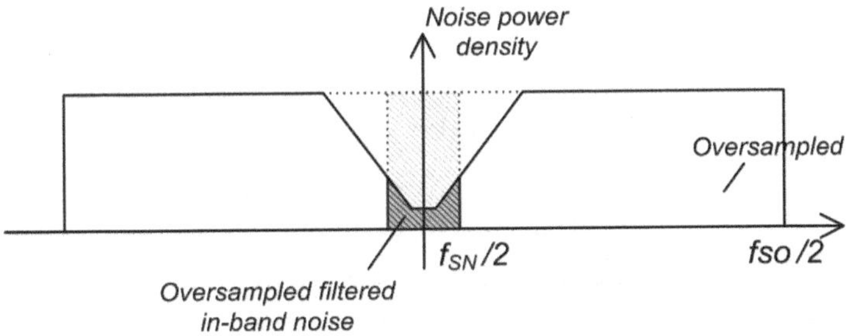

Fig. 2.22 Noise power density at the output of a classic ΔΣ modulator

specifications in terms of linearity/efficiency or noise sensitivity/consumption. For that purpose, the detection of the addressed power must be carried out at the expense of low die area consumption, and with some degree of transparency on the *RF* operation. In literature, *Built-In Current Sensor* (*BICS*) are commonly used to transparently probe the consumed current that is drawn on power supplies by digital or analog *Devices Under Test* or *DUT* (Vázquez and Pineda de Gyvez 2004). Failure detection is then made possible by monitoring discrepancies in the quiescent current I_{DDQ}, and circuits presenting defaults can then be replaced by alternate back-up circuits. The proposed circuit topology is a digital *Built-In Current Sensor* whose approach is a generalization of I_{DDQ} test to I_{DDT} control where T stands for transient (Leyssenne et al. 2008, 2009b). The goal is no longer to detect current discrepancies but to dynamically provide a digitalized image of the consumed current that is drawn on a power supply. Current consumption generally increases as a function of power level in power amplifiers (or any device under test) due to 2nd-order non linearities. Consequently, the proposed *BICS* behaves as a transparent digital envelope-tracker.

Moreover, in order to address the issue of quantization noise/distortion that was previously analyzed (see Section 2.3), the *BICS* has to carry out ΔΣ modulation and provides a digital word that allows configuring the *DUT*, i.e. a parallelized multi-cell power amplifier. In the following sections, the topology and the theory underlying in the *BICS* will be presented as well as the design methodology according to the targeted application. In Section 2.4.3, a combined architecture including a *BICS* and a *WLAN* power amplifier will demonstrate the interest of envelope power control for efficiency and protection purpose.

2.4.2.2 General Topology of Delta-Sigma Built-In Current Sensor

The general schematic of ΔΣ-*BICS* is depicted in Fig. 2.23. It is essentially a closed-loop circuit based on *Maidon* topology (Maidon et al. 2000). The main difference lies in the discretized feedback current interface. This *BICS* is made of four distinct blocks: an input mirror, a noise shaping filter/amplifier, a multi-bit voltage comparator, and a multi-bit current digital-to-analogue converter. The 2^M output digital signals are noted $\Phi_{ON}(j)$.

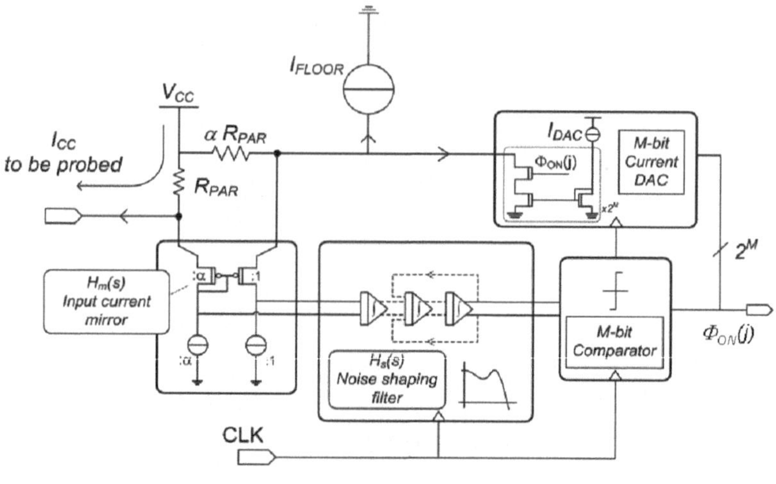

Fig. 2.23 ΔΣ *BICS* synoptic

When the consumed current increases, it generates a voltage drop through a parasitic resistor R_{par} which unbalances the input mirror. This results in a voltage deviation that is proportional to the input current (*Delta* operation) and that is amplified and filtered by the transfer function $H_s(s)$ (*Sigma* operation). The error voltage is then converted into a digital word by means of comparators. The current *DACs* achieve the current feedback that is necessary to compensate for the voltage unbalance at the input mirror and to guaranty a stable closed-loop behavior. This topology can be considered as a ΔΣ Analog-to-digital converter to the extent quantization noise is rejected out of the frequency bandwidth of interest. The block diagram can be simplified in the following form (Fig. 2.24):

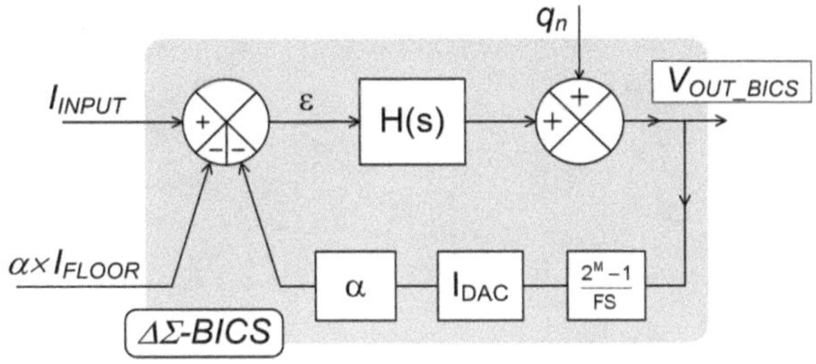

Fig. 2.24 ΔΣ *BICS* block diagram

where M is the resolution of the topology. FS represents the full-scale output voltage range over which the *BICS* operates. I_{FLOOR} represents the output-referred floor of the operating current range. I_{DAC} represents the output current of a single current *DAC*. The factor α represents the *BICS* current dividing ratio and q_n represents the quantization voltage noise. H is the trans-impedance transfer function:

$$H(s) = H_m(s) \times H_s(s) \tag{2.14}$$

where $H_m(s)$ and $H_s(s)$ are respectively the input mirror transfer function and the noise shaping transfer function.

The output voltage of the *BICS* (i.e. the arithmetic mean of the comparator output digital signals) is given by the following relationship including quantization noise q_n:

$$V_{BICS}(s) = (I_{INPUT}(s) - \alpha \cdot I_{FLOOR}) \cdot STF(s) + q_n \cdot NTF(s) \tag{2.15}$$

where

$$STF(s) = \frac{H(s)}{\left(1 + \frac{\alpha \cdot 2^M \cdot I_{DAC} \cdot H(s)}{FS}\right)}$$

$$NTF(s) = \frac{1}{\left(1 + \frac{\alpha \cdot 2^M \cdot I_{DAC} \cdot H(s)}{FS}\right)}$$

Limit conditions therefore allow finding the appropriate values for I_{FLOOR} and I_{DAC} as a function of the targeted current range $\Delta I_{INPUT\_range} = [I_{NPUT\_min}; I_{NPUT\_max}]$:

$$V_{BICS} = 0|_{@I_{INPUT}=I_{INPUT\_min}} \Leftrightarrow I_{FLOOR} = \frac{I_{INPUT\_min}}{\alpha} \tag{2.16}$$

$$V_{BICS} = FS|_{@I_{INPUT}=I_{INPUT\_max}} \Leftrightarrow I_{DAC} = \frac{(I_{INPUT\_max} - I_{INPUT\_min})}{\alpha \cdot 2^M} \tag{2.17}$$

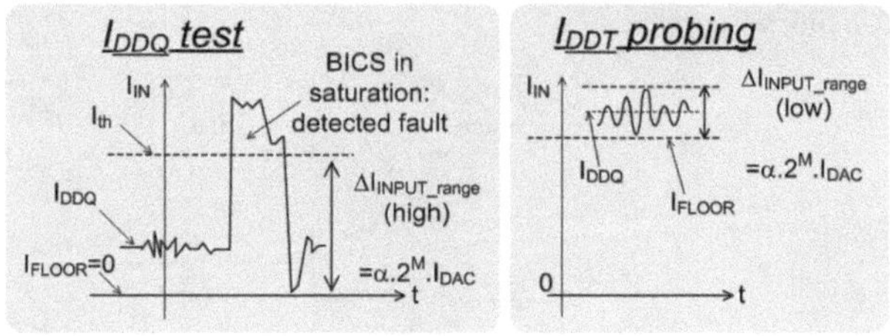

Fig. 2.25 Illustration of I_{DDQ} test and I_{DDT} control

According whether I_{DDQ} test or I_{DDT} probing is targeted, I_{FLOOR} and I_{DAC} should be reconfigured as illustrated in Fig. 2.25. The factor α should be chosen according to the trade-off between consumption, die-area consumption, and mismatch sensitivity (high α results in high sensitivity and die area).

Assuming H is wide-band enough to cover the targeted bandwidth, the theoretical *RMS* voltage noise level over this bandwidth can be derived according to (2.15) and is given by:

$$\sqrt{\overline{v_n^2}} = \frac{FS^2}{2^{2M+1} \cdot \alpha \cdot I_{DAC} \cdot |H(0)| \cdot \sqrt{3 \cdot OSR}} \tag{2.18}$$

Identically, the theoretical input-referred *RMS* current noise level over this bandwidth is given by:

$$\sqrt{\overline{i_n^2}} = \frac{FS}{2^{M+1} \cdot |H(0)| \cdot \sqrt{3 \cdot OSR}} \tag{2.19}$$

2.4.2.3 Example of a Low-Pass 1st Order 200 kHz Delta-Sigma Built-In Current Sensor

In the following example, a narrow-band ΔΣ *BICS* devoted to *RSSI* (Received Signal Strength Indicator) for *GSM/GPRS/EDGE* application is proposed as an alternative to (Khoo 1998, Huang et al. 2000). In this case, the ΔΣ *BICS* allows probing the current that is consumed by the receiver blocks working in the base-band frequency domain.

The targeted full-scale current swing is 2 mA. In order to limit die area, the *BICS* dividing ratio is chosen as high as $\alpha = 10$. The parasitic resistor was reduced down to 1.5 Ω. The full scale voltage range is chosen close to $V_{DD}/2 \sim 1.5$ V ($V_{DD} = 2.5$ V since a 0.25 μm *BICMOS* technology is used). In order to limit the *BICS* current consumption (~4/5 mA), the noise shaping filter is made as simple as possible by means of a single order switched capacitor differential integrator as follows:

$$H(s) = H_m(0) \times \frac{H_S(0)}{\left(1 + j\frac{f}{f_0}\right)} \tag{2.20}$$

where $H_m(0) = 25\ \Omega$, $H_s(0) = 200$ and $f_0 = 1$ MHz

The resolution, the over-sampling ratio and the clock frequency were chosen to be respectively $M = 4$, $OSR = 160$ and $f_s = 64$ MHz so as to keep the input-referred *RMS* current noise below 1 μA. Figure 2.26 depicts the *BICS* output voltage (arithmetic mean of digital control bits) both in frequency and time domain when a 200 kHz sine wave input current is applied.

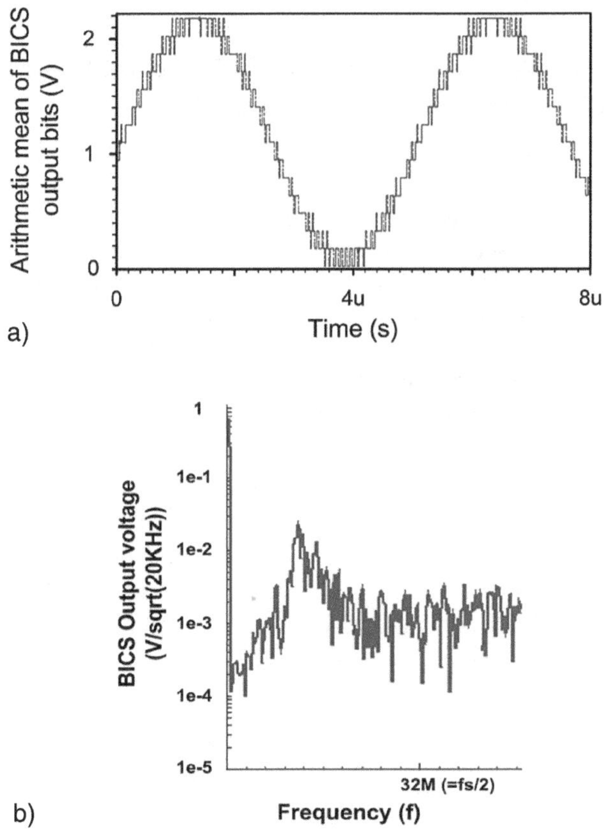

Fig. 2.26 ΔΣ-*BICS* dual-tone output illustration, (**a**) in time domain, (**b**) in frequency domain when a 200 kHz 2 mA peak-to-peak current is probed (50 μs simulation, Blackman window)

The quantization noise density in the targeted bandwidth is 1.4 μV/sqrt(Hz) (i.e. 0.2 mV/sqrt(20 kHz) as depicted on Fig. 2.26). The *RMS* noise level over a 200 kHz bandwidth is therefore given by:

$$\sqrt{\overline{v_n^2}} = \sqrt{\int_0^{200KHz} 1.4 \cdot 10^{-6} df} = 0.6mV_{RMS}$$

In an analog manner, the input referred *RMS* current noise over 200 kHz is found to be 0.8 μA. Dynamic range over the 200 kHz band is given by:

$$DR = 10\log\left(\frac{\overline{v_{signal}^2}}{\overline{v_n^2}}\right) = 10\log\left(\frac{FS^2}{8 \cdot \overline{v_n^2}}\right) \approx 59\,\text{dB}$$

2.4.3 Dynamically Reconfigurable RF Power Amplifier Controlled via Delta-Sigma Built-In Current Sensor

2.4.3.1 Design Methodology

A ΔΣ *BICS* can also be used to control an *RF* power amplifier in terms of linearity/efficiency, and the schematic of such adaptively reconfigurable architecture is depicted in Fig. 2.27. The adaptive bias mechanism consists in dynamically and adaptively reconfiguring the number of activated cells according to the *RF* envelope (Leyssenne et al. 2009b). When *RF* power is applied to the *PA*(①), the 2nd order non-linearity of the input admittance generates an envelope current, I_{ENV}(②), that propagates through the operational amplifier buffers. The ΔΣ *BICS* allows probing

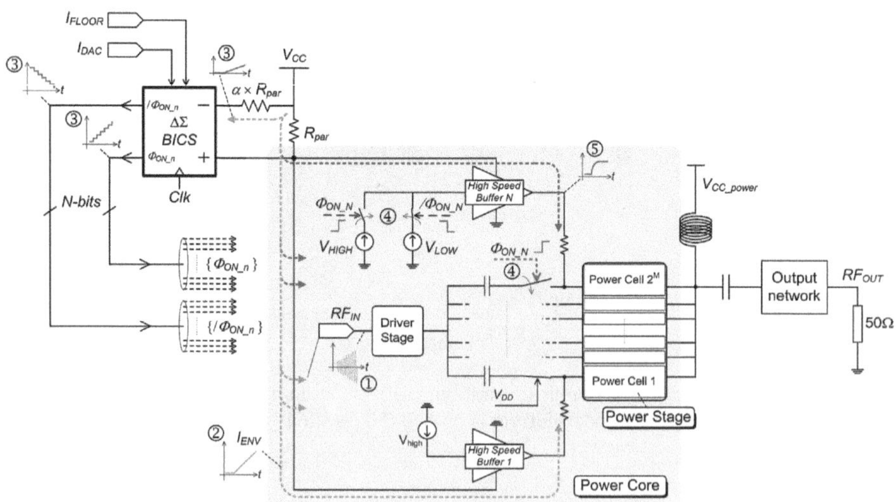

Fig. 2.27 Adaptively reconfigurable *PA* architecture based on a ΔΣ *BICS* and a *PA* core to be controlled

this extra current in a transparent manner and provides a control word (③) that reconfigures the *PA* core accordingly (④) and biases the appropriate number of power cells ⑤ with a rising time period of 1 ns.

Envelope variations (characterized by peak-to-average ratio or Cubic metric) determine the reconfiguration power range, i.e. the architecture resolution and *BICS* parameters I_{DAC} and I_{FLOOR}. For sake of low complexity, resolution was limited to $M = 3$. In order to relax the output filter selectivity (and consequently its in-band losses), *BICS* bandwidth is designed high enough to provide control over the whole *WLAN* band.

2.4.3.2 Theoretical Approach of Power Detection in a *PA* Control Architecture

The relationship between the envelope current I_{ENV} and the input *RF* envelope depends on the number of active cells and can be roughly approximated by the following linear interpolation:

$$I_{ENV} = \left(Y_{\pi 2,1} - w_{ctrl}\left(t\right) \times \Delta Y_{\pi 2}\right) \times X_{IN}\left(t\right) \tag{2.21}$$

where X_{IN} is the squared input *RF* envelope ($X_{IN} = V^2{}_{IN}$),
$w_{ctrl}(t) = V_{BICS}(t)/FS$ is the normalized digital control word,
$Y_{\pi 2,j}$ is the 2nd order input admittance when *j* power cells are activated,
$\Delta Y_{\pi 2} = (Y_{\pi 2,1} - Y_{\pi 2,2M})$ is the maximum deviation in the 2nd order input admittance.

In the case of a 3-bit *LDMOS* multi-cell power amplifier, the deviation of the probed current I_{ENV} with respect to input power is given in Fig. 2.28 as a function of the number of active cells. The maximum value for the deviation ΔI_{ENV} is in the order of 2 mA. Therefore, such dependency introduces a (non-linear) negative feedback in the closed-loop behavior.

Compared with Fig. 2.24, the block diagram must be slightly updated with an extra feedback loop as depicted in the dark region of Fig. 2.29. Moreover, when bipolar transistors are used in the power core, the quiescent base current I_π must be included in the overall current contribution (depicted by the dotted line feedback in Fig. 2.29).

In addition, the system must warranty that there is at least one active power cell at any time. This results in a slight decrease of the effective resolution, and the full current range is reduced to $(2^M - 1) \times I_{DAC}$.

Assuming *H* is wide-band enough to cover the targeted bandwidth ($H(s) = H_0$ in the band), the *BICS* output can be related to the input envelope according to the following non-linear expression:

$$V_{BICS} = \frac{q_{n,w} + H_0\, Y_{\pi 2,1} \times \left(X_{IN} - X_{IN\_\min}\right)}{\left(1 + \frac{H_0}{FS} \times \left(\Delta Y_{\pi 2} \times \left(X_{IN} - X_{IN\_\min}\right) + Y_{\pi 2,N} \times \left(X_{IN\_\max} - X_{IN\_\min}\right)\right)\right)} \tag{2.22}$$

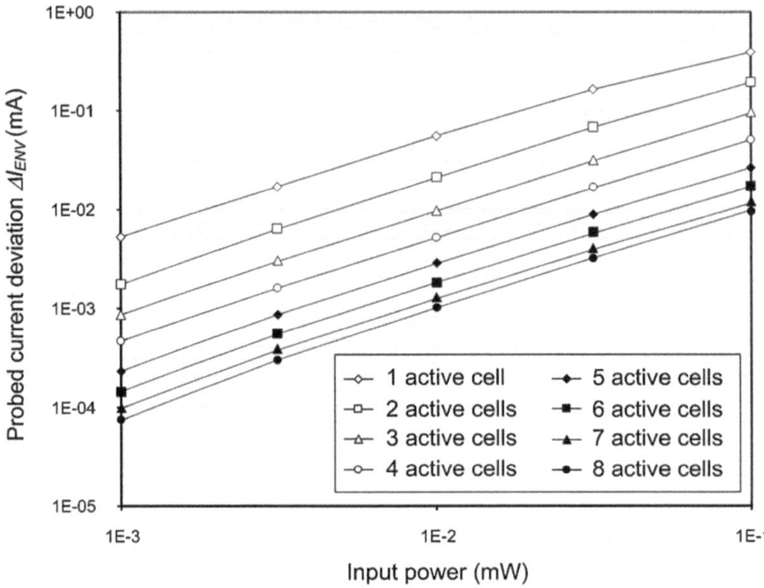

Fig. 2.28 Probed current deviation vs. input power as a function of the number of active power cells

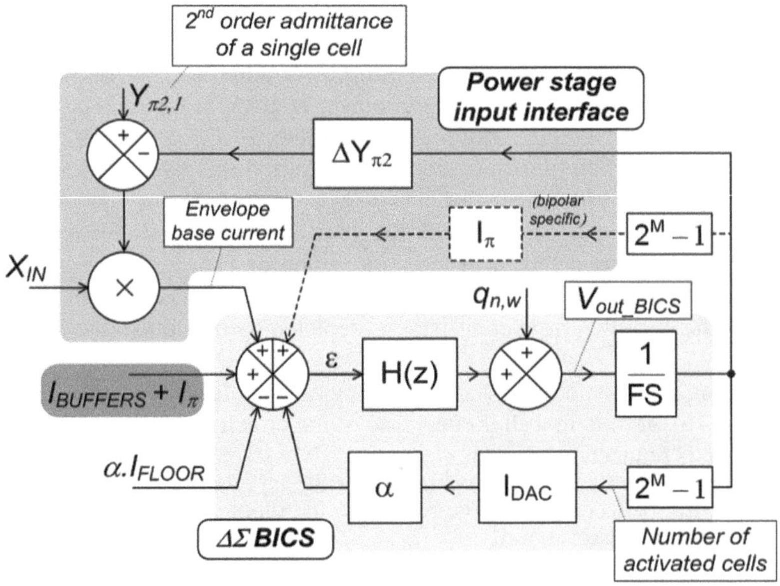

Fig. 2.29 Illustration of I_{DDT} control principle in the case of an *RF* power stage dynamic reconfiguration

where X_{IN_min} (resp. X_{IN_max}) is the minimum (resp. maximum) input voltage envelope, and $q_{n,w}$ is the quantization white noise.

The envelope borders X_{IN_min} and X_{IN_max} are related to the *BICS* parameters as follows:

$$X_{IN\_min} = \frac{\alpha \cdot I_{FLOOR} - I_{BUFFER} - I_{\pi}}{Y_{\pi 2,1}} \tag{2.23}$$

and

$$\alpha \cdot I_{DAC} - I_{\pi} = \frac{Y_{\pi 2,N} \cdot X_{IN\_max} - Y_{\pi 2,1} \cdot X_{IN\_min}}{2^{M} - 1} \tag{2.24}$$

In this case, the current I_{DAC} will determine the power range over which reconfigurability is considered whereas I_{FLOOR} will determine its bottom corner. Those parameters therefore play a key role and must be optimized according to the targeted application (maximum specified power, peak to average ratio...). An increase in the resolution M for a more linear control of the *PA* will therefore necessitate an improved accuracy of I_{DAC} or a reduction in the ratio α. Practically, I_{DAC} and I_{FLOOR} currents are in the order of dozens of µA. Special care must be paid to the design of the sources that generate them. Monte Carlo analyses are required to determine the sensitivity of linearity to these current sources if they are integrated on-chip. Trimming or external calibration may also be considered.

Equation (2.22) presents some analogies with (2.15) but its non-linear behavior with regard to the input envelope X_{IN} implies that the DC power consumption (i.e. the number of active cells) is overestimated at low power levels and efficiency is not fully optimized. The power sub-ranges are therefore not homogeneous as depicted in Fig. 2.30. Moreover, (2.22) suggests that quantization white noise is convoluted with the input envelope, thus resulting in a power-dependent pink noise in the vicinity of DC. This pink noise degrades the spectral purity of the output voltage.

In order to combat non-ideal effects in the power detection response (i.e. thermal or non-linear behavior as enlightened by (2.22)), a *BICS* linearization technique consist in introducing a compensation non-linearity by means of non-homogeneous current *DACs*. E.g., the magnitude of the 2nd-order input admittance $Y_{\pi,2}$ is all the lower as the number of active cells is high (see Fig. 2.28). Consequently, *DAC* current sources should decrease consistently and gradually according to an optimization function F_{LIN}.

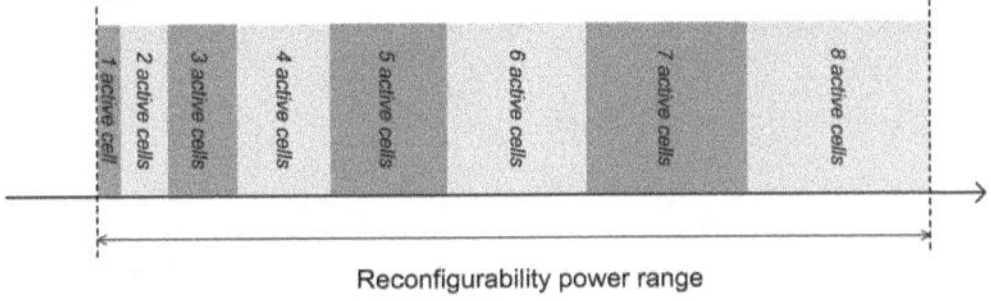

Fig. 2.30 Illustration of non homogeneous power sub-ranges

$$I_{DAC,\,i} = I_{DAC} \times \left(\frac{2^M \times F_{LIN}(i)}{\sum_{i=1}^{2^M} F_{LIN}(i)} \right) \tag{2.25}$$

where $I_{DAC,\,i}$ is the ith indexed current *DAC*.

For a 3-bit architecture, the appropriate non-homogeneous current *DAC* profile that allows linearizing the *BICS* output is represented in Fig. 2.31a, Fig. 2.31b represents the arithmetic mean of the *BICS* output bits and illustrates the linearization

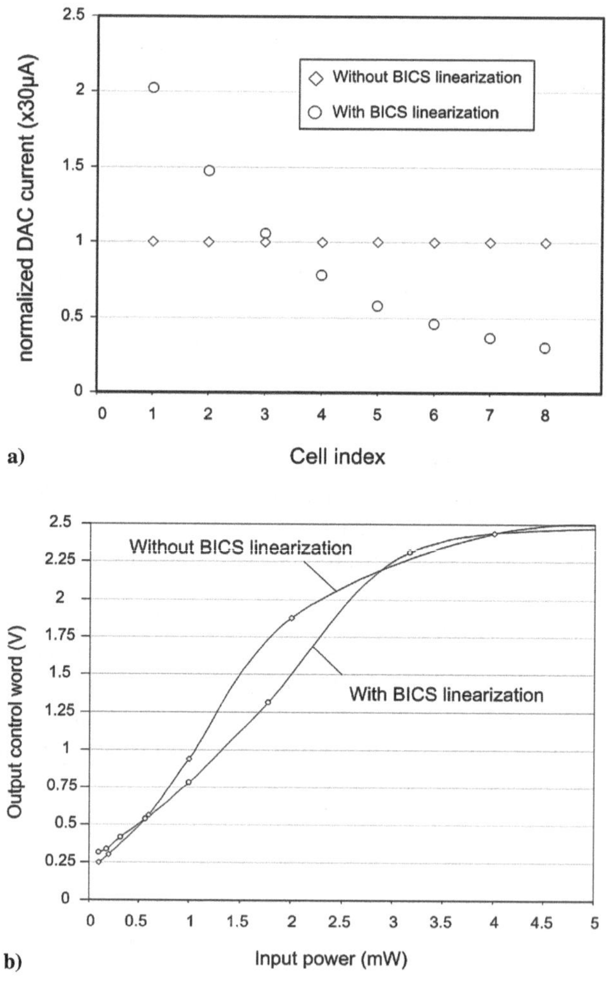

a)

b)

Fig. 2.31 Illustration of *BICS* linearization when embedded in a *PA* control architecture: (**a**) linearization *DAC* current profile, (**b**) *BICS* output digital word w/ or w/o linearization resulting from the non-homogeneous *DAC* currents when a *CW RF* signal is applied to the power core. A comparison with the non-linearized topology is also given.

Such technique allows reducing power consumption by more than 10% in the middle of the reconfiguration range. The main drawback lies in the accuracy of *DAC* current sources that must be controlled in the order of a few microamperes.

2.4.3.3 Design of the Noise Shaping Transfer Function *H*(*s*)

The applications that will be hereunder considered are *WLAN*/*WIMAX* standards for which channel bandwidth is broad (10–20 MHz) and overall operation bandwidth is 80 MHz. The noise shaping transfer function must be designed accordingly. In order to address this issue several filter topologies can be proposed. It can be desirable to introduce one (or several) transmission zero(s) in the *Noise Transfer Function* (Fig. 2.32) by means of resonant dual complex pole(s) in $H(s)$ ($\pm jB$ where B is the channel bandwidth).

$$H(s) = \frac{1}{\left(\tau_0\, s + \frac{1}{A_0}\right) \cdot \left(\left(\tau_1\, s + \frac{1}{A_0}\right)^2 + (B\tau_1)^2\right)} \underbrace{\approx}_{A_0 >> \frac{1}{\tau_0\, s}} \frac{1}{\tau_0\, s \cdot \tau_1^2 \cdot (s^2 + B^2)} \tag{2.26}$$

where A_0 is the maximum voltage gain of integrators.

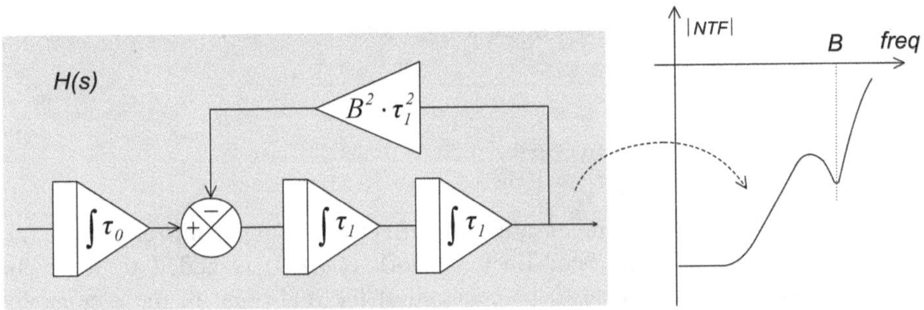

Fig. 2.32 A dual-complex-pole filter topology *H*(*s*) and the resulting noise transfer function

Meanwhile, this results in reduced phase margin and higher complexity. Another filter topology based on (Proportional/Integral/Derivator) was finally considered as depicted in Fig. 2.33. Its transfer function is in the form:

$$H(s) = \frac{1}{2} \cdot \frac{A(s)}{1 + A(s) \cdot B(s)} = \frac{1}{2} \cdot \frac{A_0 \cdot (1 + \tau_1\, s)}{A_0\, B_0 + (\tau_0 + \tau_1) \cdot s + \tau_0\, \tau_1\, s^2} \tag{2.27}$$

where $A(s) = \frac{A_0}{1+\tau_0 \cdot s}$ and $B(s) = \frac{B_0}{1+\tau_1 \cdot s}$ are respectively the transfer functions of the forward and feedback paths.

The term ½ stands for the fact that only half of the output differential signal is actually used. B_0 is kept low (~10.e–3 <1) in order to preserve the overall voltage gain. The time constants are designed as follows:

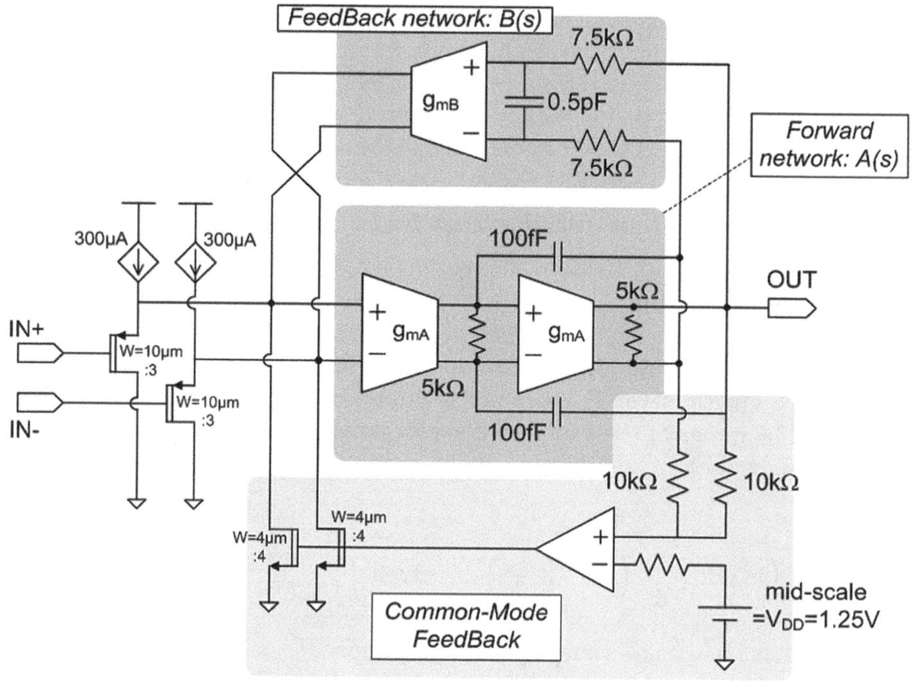

Fig. 2.33 A wide-band filter topology *H*(*s*) featuring *CMFB*

$$\begin{cases} \tau_0 \approx 4\,\text{ns} \rightarrow f_A \approx 40\,\text{MHz} \\ \tau_1 \approx 10\,\text{ns} \rightarrow f_B \approx 16\,\text{MHz} \end{cases}$$

The derivation allows compensating the effect of the input mirror cut-off frequency. A Common-Mode Feed-Back network (*CMFB*) is added to force the common-mode output voltage to be centered at the mid-point of the comparator operation range, i.e. $V_{DD}/2 = 1.25$ V. This ensures that the comparator output is an image of the input mirror current deviation and is independent of any parasitic offset voltage due to temperature, power supply variations, process mismatch.

The blocks $A(s)$ and $B(s)$ are implemented by means of simple *CMOS OTA* stages (Operational Transconductance amplifiers) whose schematic is given in Fig. 2.34. The voltage gain of *OTA* stages in $A(s)$ was maximized by choosing $I_{TAIL} = 150\,\mu\text{m}$.

2.4.3.4 Management of Linearity/Efficiency Trade-Off via a 1-Bit Delta-Sigma *BICS*-Controlled Architecture

In this section, the architecture is configured as follows. The $\Delta\Sigma$ *BICS* is a 1-bit sensor, i.e. it features only one comparator and one current *DAC*. The current I_{DAC}

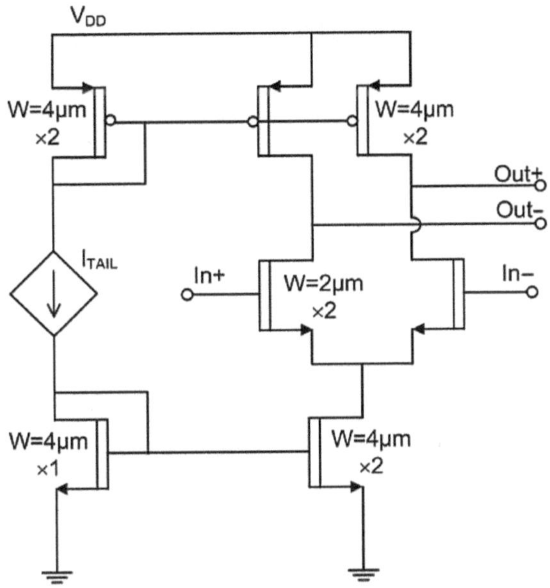

Fig. 2.34 Schematic of a *CMOS* operational transconductance amplifier stage

is set to 200 μA. If the output *BICS* is "0", only two cells of the power core are active. If the output *BICS* is "1", all cells of the power core are active. The clock frequency is set to 160 MHz. The *CW* response of such architecture is shown in Fig. 2.35.

By setting the floor current I_{FLOOR} properly (i.e. by shifting the reconfiguration power range), it is possible to minimize *AM/AM* while achieving a significant *PAE* improvement compared with a class-A power amplifier. Figure 2.35a shows that *PAE* peaking can be obtained at low power levels with this technique by setting I_{FLOOR} to 1.85 mA. This is analog to what is generally observed with a Doherty architecture. Meanwhile this is at the expense of an excessive modulation of power gain (Fig. 2.35b) at high power levels. By setting I_{FLOOR} to 1.825 mA, the variations of power gain as a function of input power (*AM/AM*) are strongly mitigated. It is therefore possible to optimize both efficiency and linearity by a proper interpolation between these two settings providing I_{FLOOR} is controlled with an accuracy in the order of a few dozens of μA.

When a 20 MHz-wide *WLAN* signal (*QPSK OFDM* modulation) is applied to the 1-bit reconfigurable *PA* (the clock frequency is still set to 160 MHz, *OSR*=4), the output *RF* envelope follows the *RF* input one except for the noise (Fig. 2.36).

From a spectral point of view (Fig. 2.37), the *BICS* output (arithmetic mean) and the *RF* output magnitude presents good noise response in the vicinity of the channel but poorer behavior out-of-band. Such high quantization noise results in a *RMS EVM* up to 16%, which is not acceptable. In order to circumvent this issue, it is necessary to increase the system resolution and bandwidth. In the following development, a 3-bit architecture is considered.

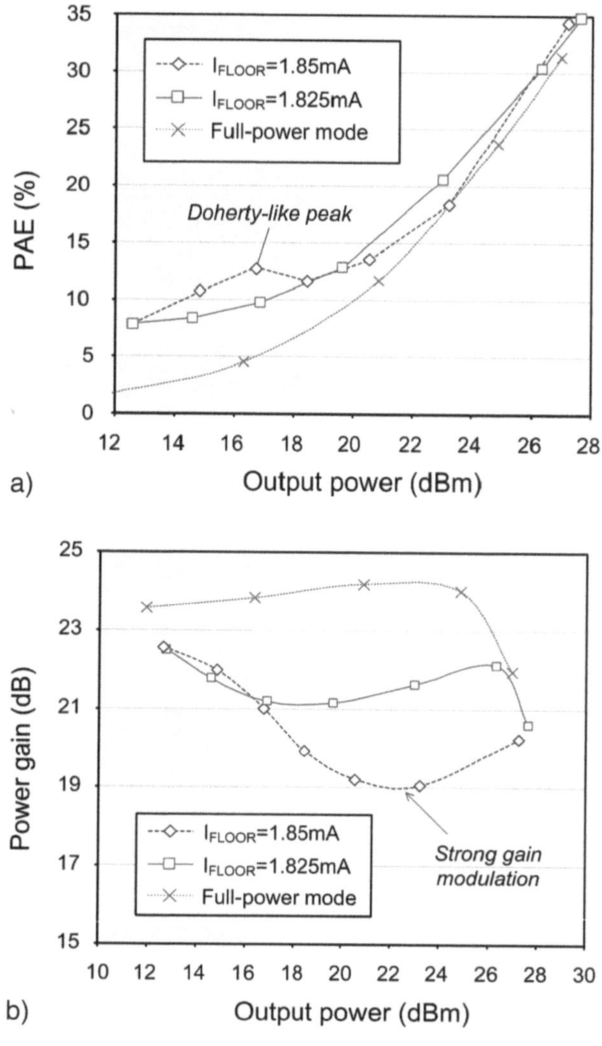

Fig. 2.35 (**a**) *PAE* and (**b**) power gain of a 1bit-ΔΣ *BICS*-controlled *PA*

2.4.3.5 Management of Linearity/Efficiency Trade-Off via 3-Bit Delta-Sigma *BICS*-Controlled Architecture

In this section, the architecture is configured as follows. The ΔΣ-*BICS* is a 3-bit sensor, i.e. it features 8 comparators/current *DACs*. The current I_{DAC} is set to a 30 μA fixed value and the output *BICS* linearization technique presented in Section 2.4.3.5 is employed. The clock frequency is still set to 160 MHz. The *RF* power core is loaded by a generic Butterworth *RF* output filter. In order to enlighten the benefit of this architecture on efficiency, this filter is set lossless and presents a 80 MHz bandwidth with a 20 dB rejection at 50 MHz from the center frequency. Hereunder,

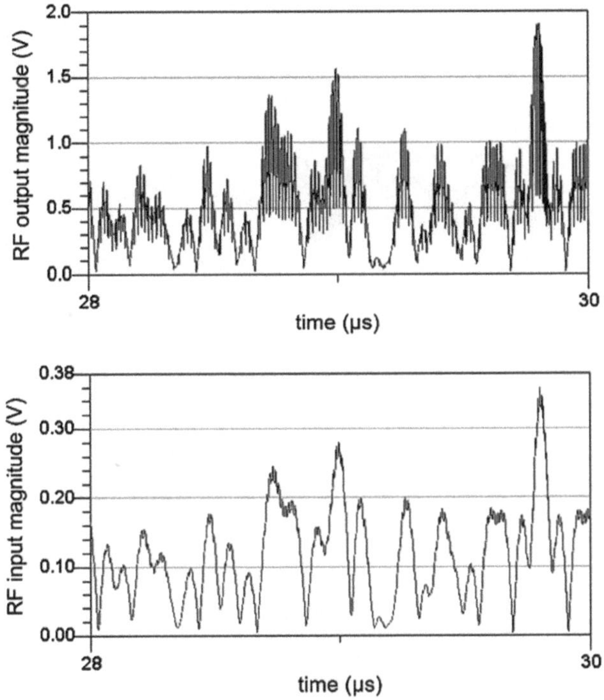

Fig. 2.36 Input and output envelopes when a *WLAN* signal is applied to a 1-bit power adaptability (time domain)

the *PA* architecture performance is analyzed by applying a 1 MHz-spaced dual tone signal. In order to reduce the envelope dynamic, one of the tones is made four times as powerful as its counterpart. Dotted lines represent the *PA* response if it was purely linear.

As expected, the output *RF* magnitude tracks the ideal one with alternate up and down steps (Fig. 2.38a). The quantization noise is rejected out of the useful bandwidth and is greatly cancelled by the output band-pass filter (Fig. 2.38b). At very high power level ($P_{IN} = 10$ mW), the adaptive system *saturates*, the *PA* reaches its ultimate power capability and linearity is degraded. At the opposite corner of the *BICS* detection power range ($P_{IN} = 0.3$ mW), some linearity decrease is also observed. This is due to the relative magnitude steps that are higher at lower power levels. This results in significant magnitude glitches in the filtered response (Fig. 2.38b). Indeed, quantization noise is not equally distributed over the reconfiguration range which is an important difference with $\Delta\Sigma$ *ADC*. In the middle of the power reconfiguration range, the magnitude response is optimal (with minor residual noise). Figure 2.38c represents the arithmetic mean of the *BICS* digital control bits. Figure 2.38d illustrates the power-dependent step-varying phase shift. The maximum phase deviation is kept below 15°. Figure 2.38e stresses the power–dependent current consumption control over a 6:1 ratio.

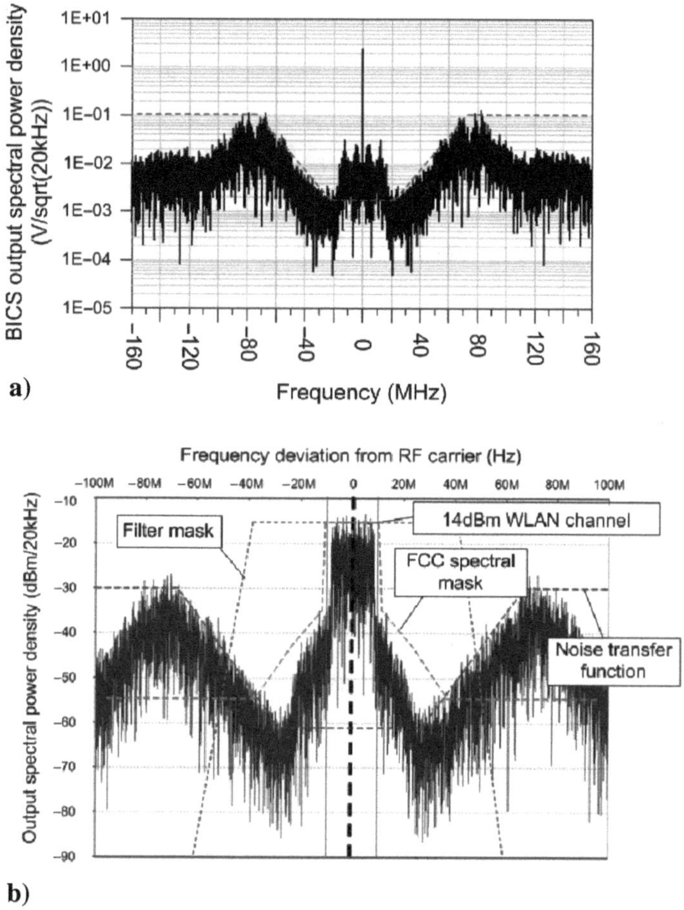

Fig. 2.37 (**a**) 1-bit *BICS* output, (**b**) *RF* output voltage in frequency domain when a *WLAN RF* signal is applied to a 1-bit architecture

A 20 MHz-wide *WLAN* signal (*QPSK OFDM* modulation) is then applied to the 3-bit reconfigurable *PA* (the clock frequency is still set to 160 MHz, *OSR*=4). Figure 2.39 depicts the chronograms of the *BICS* output voltage (arithmetic mean of control digital signals), the *RF* input/output envelopes (output is considered after band-pass *RF* filtering), and the dynamic consumed current. As expected, the *BICS* output and the dynamic consumed current properly track the *RF* envelope variations.

Figure 2.40b shows that the *FCC* spectral mask is fulfilled in the vicinity of the channel. However, discrepancies can be observed at 40 MHz from the carrier where the –40 dBc spurious rejection is hardly fulfilled. This stresses to what extent the loop bandwidth limitation impacts the output *SNDR*. Special care must be given to the design of the filter $H_S(f)$ that must be wide-band enough while preserving

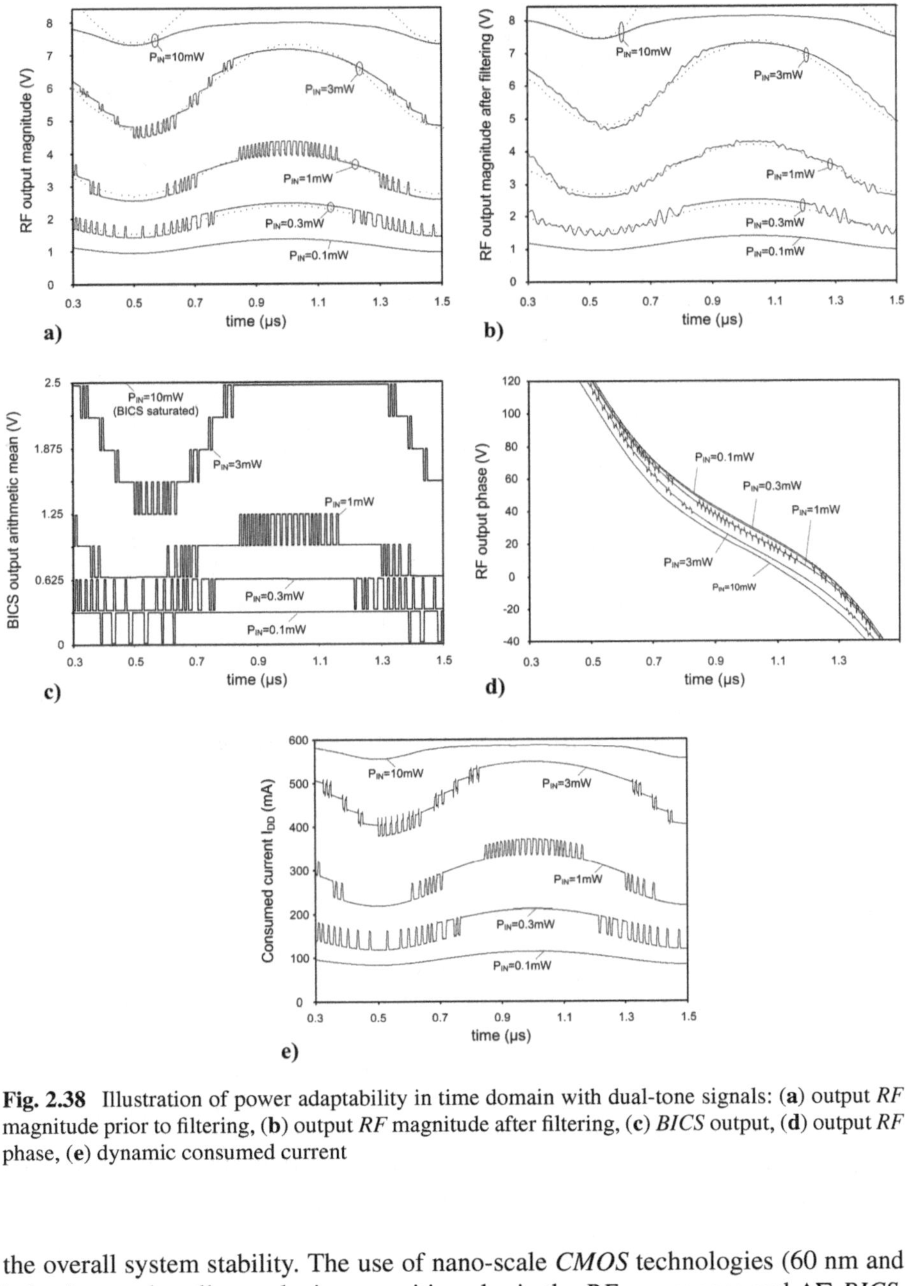

Fig. 2.38 Illustration of power adaptability in time domain with dual-tone signals: (**a**) output *RF* magnitude prior to filtering, (**b**) output *RF* magnitude after filtering, (**c**) *BICS* output, (**d**) output *RF* phase, (**e**) dynamic consumed current

the overall system stability. The use of nano-scale *CMOS* technologies (60 nm and below) may also allow reducing parasitic poles in the *RF* power core and ΔΣ *BICS*, thus facilitating the implementation of such architecture.

Another point to consider is that phase noise at the *PA* output is determined by the integration of the whole white noise power at *BICS* output (and not by the noise power density itself). This issue raises the question of digital filtering. Indeed,

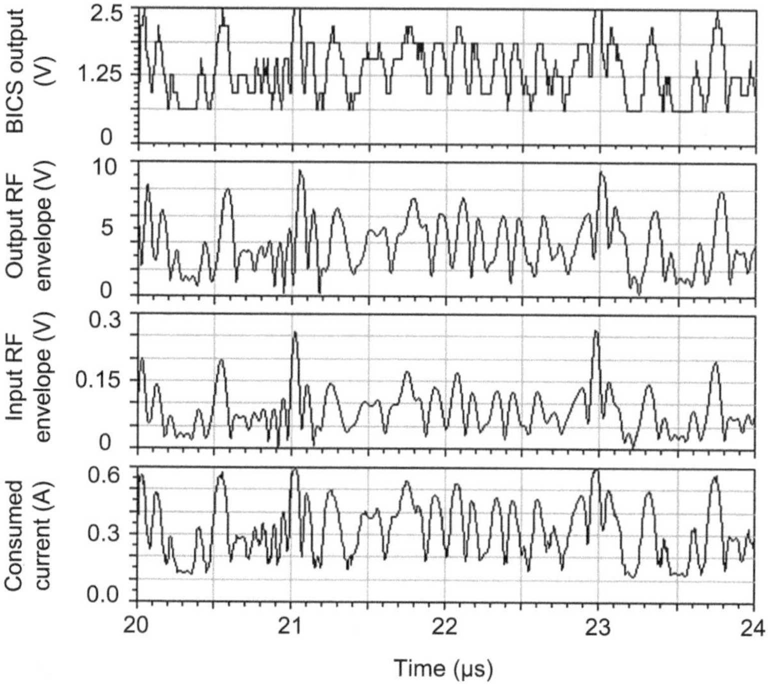

Fig. 2.39 Illustration of 3-bit power adaptability in time domain with a *WLAN* signal: (**a**) *BICS* output, (**b**) *RF* output envelope, (**c**) *RF* input envelope, (**d**) consumed current

inserting a base-band decimator digital filter between the digital *BICS* output and the control interface of the power core (Fig. 2.41) is an interesting prospect to tackle.

Such filter absorbs a substantial part of quantization white noise energy, whereby phase noise can be significantly reduced at *PA* output. The payload is greater complexity and response latency and/or stability issues. A 2nd order decimator is generally considered for a 1st order $\Delta\Sigma$ converter. This implies at least a $2/f_{CLK} =$ 12.5 ns delay time in the decimator filter which prevents the digital control word to be perfectly synchronized with *RF* envelope. Therefore, the value of f_{CLK} (i.e. the over-sampling ratio) cannot be relaxed.

Figure 2.42 compares the *Error Vector Magnitude* and the drain efficiency when the *BICS* control system is used and when all cells are constantly active (~class 0A). The collector efficiency clearly shows the benefit of this *BICS*-based-adaptive system over a class-A operation. This is to the cost of moderate linearity degradation at low power levels. The increase of *EVM* at low power levels (which echoes the linearity degradation observed in dual-tone simulation) can be explained by the fact that the *BICS* does no longer operate in the middle of its operating range but in the bottom corner which degrades its accuracy. A decrease of the equivalent over-sampling ratio is then observed thus folding noise in the bandwidth of interest.

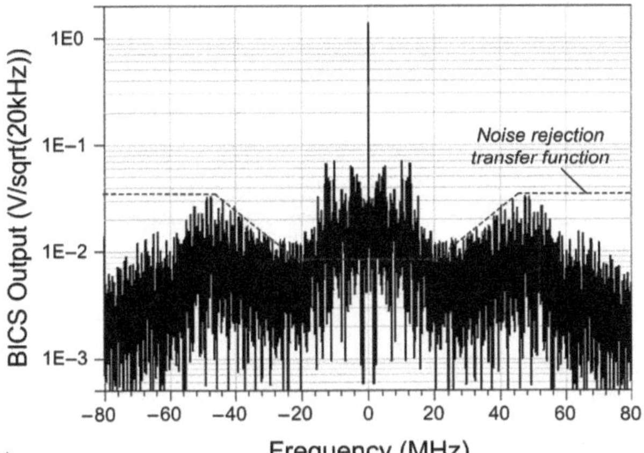

a)

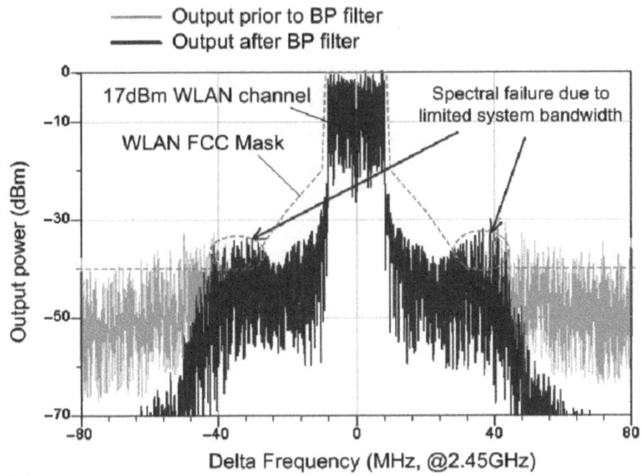

b)

Fig. 2.40 Illustration of 3-bit discretized *PA* performance in frequency domain with a *WLAN* signal: (**a**) *BICS* output, (**b**) *RF* output envelope

2.4.3.6 Impact of *VSWR* Mismatch on a Delta–Sigma *BICS*-Controlled *PA*

VSWR mismatch impacts the probed current I_{ENV}. Indeed, the more the output load, the higher the input gate capacitance and the lower the *RF* gate voltage swing. Consequently, the excess base-band current that is drawn on the activation buffers decreases with increasing output loads. The *PA* therefore gets undersized (i.e. the average number of active cells decreases) when the output load exceeds its nominal value. The *PA* is then current-curbed and protected from excessive output voltage swing which is an interesting property in terms of robustness. In an analogue manner, the *PA* gets oversized (i.e. the average number of active cells increases) when the

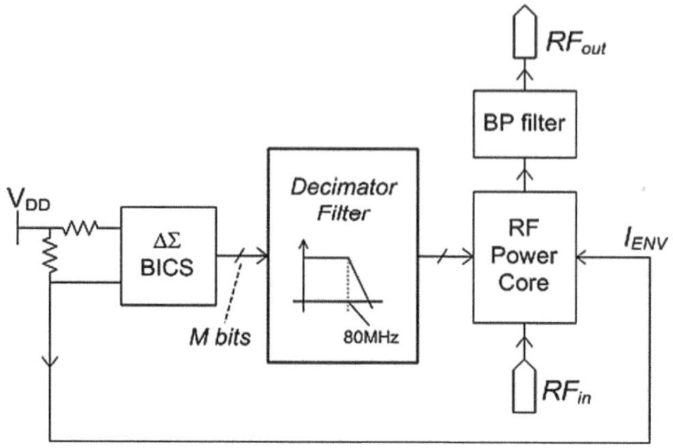

Fig. 2.41 Additional decimator filter

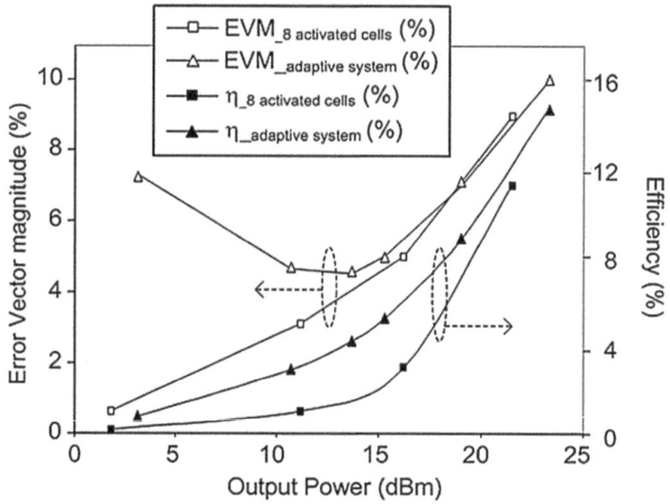

Fig. 2.42 *PA Error Vector Magnitude* and efficiency with a *WLAN* signal

output load is below its nominal value. This allows boosting the *PA* power capability and compensating the *VSWR* effect.

2.4.3.7 Conclusion on Delta–Sigma *BICS*-Controlled Power Amplifiers

One of the strengths of power amplifiers controlled by ΔΣ-*BICS* lies in the transparency of this detection principle. Indeed, A *Built-In-Current Sensor* is a low area, low consumption circuit that is capable to probe envelope deviation due to *RF* input power with no electrical impact on the *RF* power core. Its differential closed-loop topology makes it little sensitive to power supply variation (high *PSRR*), process mismatch or thermal drift. Its accuracy is mainly determined by the feedback current *DAC*, as well as its resolution and its over-sampling ratio. When used in combination with a fragmented *RF* power core, the optimization of linearity over the targeted power range is made possible providing the parameters of the *BICS* (I_{FLOOR}, I_{DAC}, F_{LIN} function, *OSR*, resolution, $H(s)$) are properly calibrated. Such parameterizability is a key feature of this architecture and calibration should be understood as standard-dependent. E.g., the current I_{DAC} must be ruled according to the dynamic range, i.e. the power range over which reconfiguration is expected (see *PAPR* or *Cubic metric*). The *BICS* noise shaping transfer function are determined by the *RF* operation bandwidth. Even though this *PA* control architecture presents *soft feedback* in the base-band frequency domain (see grey box in Fig. 2.29), it fundamentally operates as an open loop whereby the whole effort is put on the *BICS* performance. The insertion of a digital filter allows relaxing the requirements on the sensor. Meanwhile, *EVM* degradation can be encountered at low power levels. That is a reason why closed-loop alternatives are sometimes considered as it is the case in Section 2.5.

2.5 Delta-Sigma-Like Closed-Loop Dynamically Reconfigurable Power Amplifier

2.5.1 Principle

The principle of the hereunder topology is a generalization/extension of the technique proposed by (Su and McFarland 1998). The architecture behaves as a $\Delta\Sigma$*-like ADC* operating in the base-band frequency domain (envelope rate) and considers the *RF* envelope as the value to digitalize (Leyssenne et al. 2009a). The term *-like* is used because some significant differences with classic $\Delta\Sigma$ *ADC* architecture are to be noted as it will be shown below. The fundamental goal of this structure is to dynamically monitor the *Error Vector Magnitude* variations and to reconfigure the *PA* accordingly so as to reduce current consumption at the expense of strict minimum linearity degradation.

2.5.2 Architecture Synoptic, Block Diagram and Theory

2.5.2.1 Architecture Synoptic and Block Diagram

The architecture synoptic is as depicted in Fig. 2.43. It is fundamentally based on the same power core as previously described throughout Sections 2.3 and 2.4. The output filter remains of great importance to filter out the out-of-band noise/spurious that is introduced by such discretized architecture. The architecture operates as follows. The distortion of the output *RF* signal is due to non-linearity causes expressed by the *RF* power core non-linear kernels G_3, G_5.... but also to reconfiguration/quantification causes: $q_{n,1}$, $q_{n,3}$ ... that represent the noise spectral

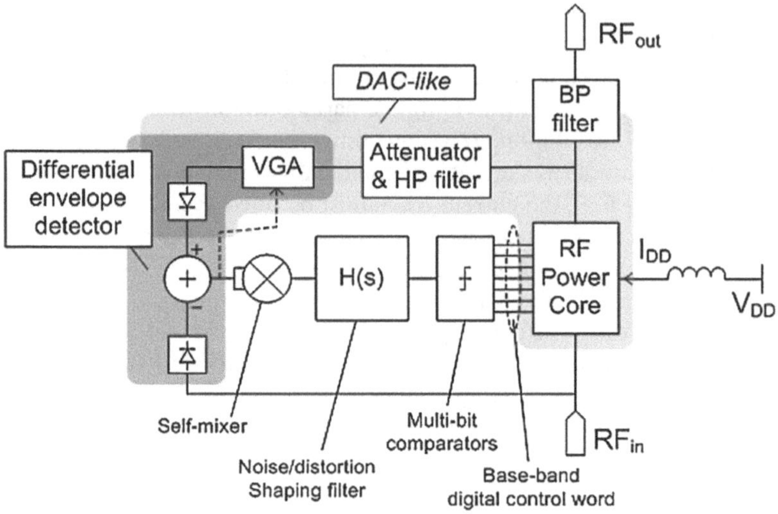

Fig. 2.43 Synoptic of ΔΣ-like closed-loop reconfigurable *PA*

densities for each *Volterra* kernel. The goal of the ΣΔ loop is to indistinctly treat both causes.

The block diagram in Fig. 2.44 shows where noise sources apply in an analog manner to a classic ΔΣ *ADC*. Output *RF* voltage is properly attenuated by means of a simple attenuator based on a RC network. This high-pass attenuator filters out the base-band glitches due to reconfiguration. The attenuation ratio is designed to match the nominal linear power gain, i.e. expressed by $1/G_1$. Input and output voltages are then converted into envelope (respectively X_{PA_IN} and X_{PA_OUT}) and differentiated (*Delta* operation) by means of a differential power detector (depicted by the dark grey zone in Fig. 2.43). The resulting signal is then squared and amplified/filtered by a base-band noise shaping function ($H(s)$). This carries out the *sigma* operation and provides a voltage V_H that is homogeneous to the squared value of the instantaneous error vector magnitude (EVM^2). By means of $H(s)$, the noise/distortion that is introduced by reconfiguration is rejected outside the operation bandwidth. If the root-mean-square of *EVM* was the value to be considered and controlled, the filter would simply consist in a single integrator that would be released every symbol period for example. However, as this circuit aims to address wide channel bandwidth applications (namely *WLAN*), instantaneous *EVM* is considered and $H(s)$ must be designed wide-band enough so that the loop closely tracks fast envelope variations. Via comparators, V_H is then converted into a digital word that activates the appropriate number of power cells. The power core is therefore adaptively "sized" according to the power level, and the non-linearity level is supposed to be kept constant over the loop operation power range.

The particularity of this ΔΣ-topology is that the role of feedback *Digital-to-Analog Converter* is played by the group {Power core; attenuator; output power

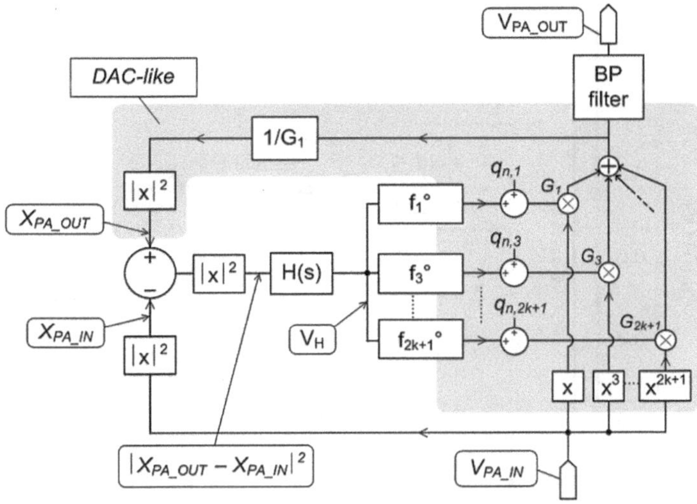

Fig. 2.44 Block diagram of ΔΣ-like closed-loop reconfigurable *PA*

detector}. This group is denoted as *DAC-like* and represented by the clear grey zone in Figs. 2.43 and 2.44. Such *DAC-like* block is non-linear contrary to classic *DAC* since it carries out successive frequency up-conversion (*RF*/digital word combination in power core) and down-conversion (envelope detection). Another particular aspect of this architecture is the insertion of additional distortion in the loop (represented by $f^{\circ}{}_1, f^{\circ}{}_3, f^{\circ}{}_5 \ldots$ in Fig. 2.44) for efficiency purpose. For these reasons, delta-sigma theory is no longer completely valid and should be amended as described in Section 2.5.2.2.

2.5.2.2 Non-linear Threshold Voltage Distribution

In order to optimize linearity/efficiency, it is desirable to introduce a correction function in the loop (that results in the functions $f^{\circ}{}_1, f^{\circ}{}_2, f^{\circ}{}_3$ in the synoptic). It consists in non-linearly distributing the comparator threshold values. Such technique echoes the non-linear *DAC* that was proposed in Section 2.4.3.2 and its implementation is limited by the same accuracy issue. These non-linearly distributed threshold voltages are depicted in Fig. 2.45 and are given by the generic expression:

$$V_{TH}(n) = V_{TH\_offset} + FS \times \left(\frac{n-1}{2^M - 1} \right)^{\gamma} \tag{2.28}$$

where $n \in [1;2^M]$ is the threshold index, *M* is the resolution, γ is an adjustment exponent, *FS* is the full-scale voltage range, V_{TH_offset} is an offset voltage that must match the common-mode output voltage of the noise-transfer function block.

The higher (resp. lower) the factor γ, the more linearity (resp. efficiency) is privileged. For high γ values, the gap between threshold voltages become tiny which results in high design difficulty and process/mismatch sensitivity.

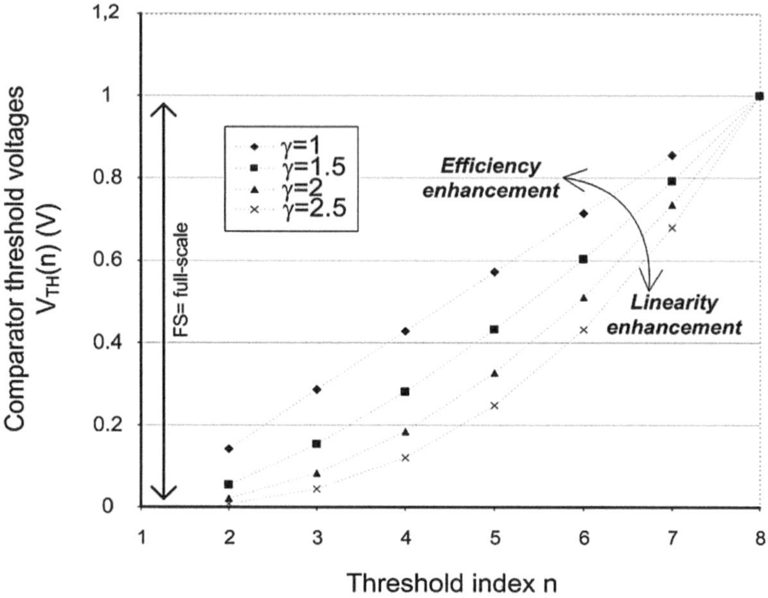

Fig. 2.45 Threshold voltage non-linear distribution

2.5.2.3 Theoretical Approach of This Architecture

The output voltage of the power core can be written in the form at third order:

$$\begin{aligned} V_{PA\_OUT} &\approx \left(G_1 + q_{n,1}\right) \cdot V_{IN} + \tfrac{3}{4} \cdot \left(G_3 + q_{n,3}\right) \cdot V_{IN}^2 \cdot V_{IN}^* \\ &\approx G_1 \cdot V_{IN} \left(1 + \varepsilon\right) \end{aligned} \tag{2.29}$$

where ε represents the relative envelope error vector.

The relative envelope error vector can be expressed by (2.30):

$$\varepsilon = \frac{q_{n,1} + \dfrac{3}{4} \cdot \left(G_3 + q_{n,3}\right) \cdot |V_{IN}|^2}{G_1} \tag{2.30}$$

where G_1 and G_3 are respectively the 1st order and 3rd order *Volterra* kernels, and $q_{n,1}$ and $q_{n,3}$ are respectively the quantization noise in 1st order and 3rd order *Volterra* kernels.

The differential envelope detector output is:

$$\begin{aligned} X_{PA\_OUT} - X_{PA\_IN} &= \tfrac{1}{2} \cdot \left|\tfrac{V_{OUT}}{G_1}\right|^2 - \tfrac{1}{2} \cdot |V_{IN}|^2 \\ &= |\varepsilon| \cdot |V_{IN}|^2 \cdot \left(\cos\left(\angle\varepsilon\right) + |\varepsilon|^2\right) \end{aligned} \tag{2.31}$$

The squarer block output is:

$$|X_{OUT} - X_{IN}|^2 = |\varepsilon|^2 \cdot |V_{IN}|^4 \cdot \left(\cos^2(\angle\varepsilon) + |\varepsilon| \cdot \cos(\angle\varepsilon) + \frac{|\varepsilon|^2}{4}\right) \tag{2.32}$$

By a proper non-linear distribution of threshold voltages, the comparator can apply a roughly square root function ($f^{\circ}{}_3(x)=x^{1/2}\ldots$) on the noise transfer function output. The third order kernel G_3 can then be built as a roughly linear function of the error vector magnitude:

$$G_3 \approx G_{3,1} + \left(G_{3,N} - G_{3,1}\right) \cdot |\varepsilon| \cdot |V_{IN}|^2 \cdot \left|\frac{H_0}{FS}\right|^{1/2} \tag{2.33}$$

where $FS = \max(|\varepsilon|^2 \cdot |V_{IN}|^4)$ is the full-scale output of the noise shaping block.

Therefore, by substituting (2.33) in (2.30), the relative error vector can be expressed in the following form:

$$\varepsilon \approx \frac{\frac{q_{n,1}}{G_1} + \frac{3}{4} \cdot \left(\frac{G_{3,1} + q_{n,3}}{G_1}\right) \cdot |V_{IN}|^2}{\left(1 + \frac{3}{4} \cdot |V_{IN}|^4 \cdot \left|\frac{H_0}{FS}\right|^{1/2} \cdot \left(\frac{G_{3,1} - G_{3,N}}{G_1 \cdot \exp(\angle\varepsilon)}\right)\right)} \tag{2.34}$$

Even though it is quite imperfect, this expression has the advantage to unify the effects of "classic" distortion ($G_{3,1}\ldots$), quantization ($q_{n,1}, q_{n,3}\ldots$), and reconfiguration enhancement ($G_{3,1} - G_{3,N}$). At low power levels, the error vector presents a positive slope with respect to power. At higher power levels, it features a flatter response and is mitigated by the reconfiguration system that is expressed by the denominator term: $|V_{IN}|^4 \times \mathrm{sqrt}(|H_0|)$. Equation (2.34) presents some analogy with classic delta-sigma modulation (see (2.15)).

2.5.3 Design Implementation

2.5.3.1 Differential Power Detector Architecture

Both detectors are made with two self-mixers whose differential outputs are cross-connected together so as to be differentiated (Fig. 2.46). If the power core is not loaded by an isolator (which is sometimes required for cost and *Bill-Of-Material* reduction purpose), power gain is sensitive to the output *VSWR* mismatch and the differential detector output is unbalanced even if no distortion is observed. To solve this issue and make this reconfigurable architecture *VSWR*-independent, it is necessary to make the attenuation adjustable in the output feedback loop. The topology that was proposed is as follows. The power detector is made controllable by means of a gain control voltage V_{GC}. An additional loop is necessary (depicted by the grey box in Fig. 2.46) to control V_{GC}.

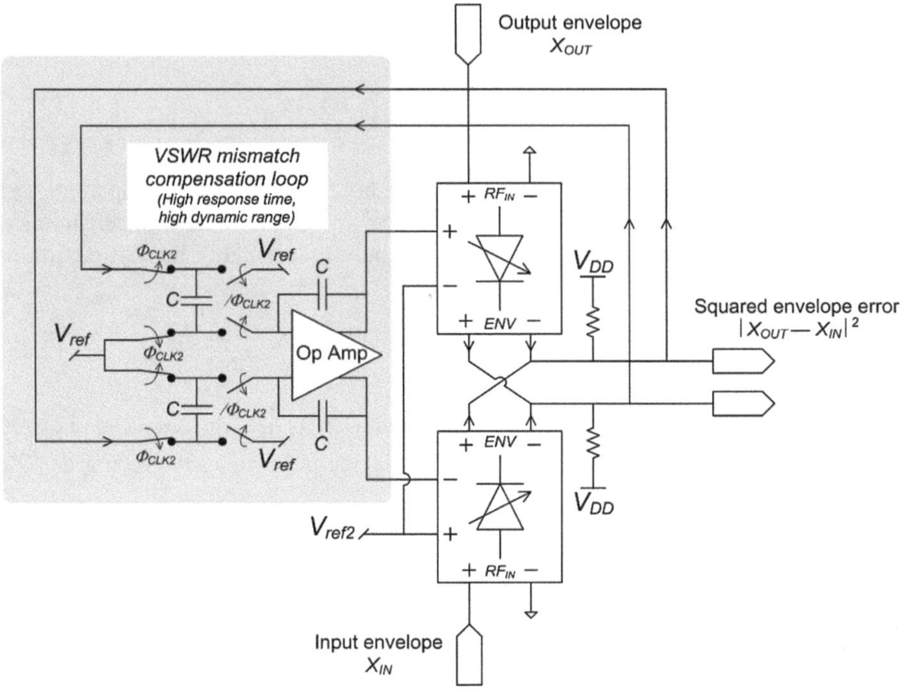

Fig. 2.46 Topology of differential power detector (Leyssenne et al. 2008)

As *VSWR* variations are very slow, this loop aims to balance the differential detector on an average basis and should behave transparently with regard to envelope-rate variations. To this end, an integrator (switched-capacitor topology is required to save die area) featuring a low frequency clock CLK_2 and a very long time-constant (in the order of time symbol or longer, i.e. 3 μs for *WLAN*) is used. If a sudden *VSWR* mismatch is applied to the *PA* core (Fig. 2.47), the squared *EVM* signal that is applied to the comparator is maximum and the digital control word *saturates* to the upper limit whatever the phase of the load coefficient Γ_L might be. This way, one ensures the *PA* linearity to remain high. The gain control voltage V_{GC} of the differential power detector slowly deviates towards a *VSWR*-dependent value. It must be remarked that V_{GC} is positive (resp. negative) for low (resp. high) output load values.

After the transient duration, input and output envelopes are balanced back again on an average basis (i.e. by averaging over a slot). In the meantime, linearity is privileged by default.

2.5.3.2 Employed Power Detector Topology and Theory

Each adjustable envelope detector is implemented with a modified *HBT* Gilbert cell whose transconductance stage is built with cross-coupled differential pairs

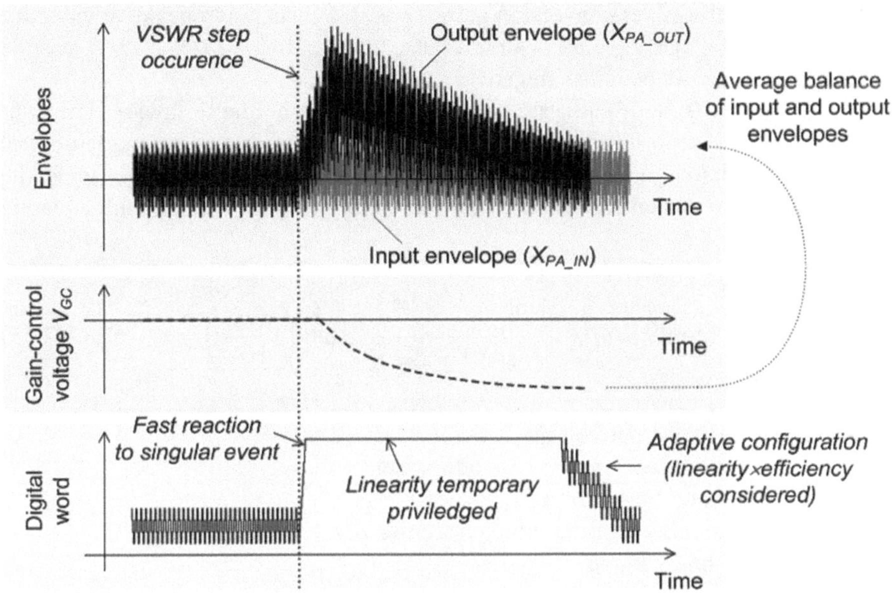

Fig. 2.47 Illustration of the slow-rate dual-loop reaction (output envelope and control signals) to a sudden *VSWR* step

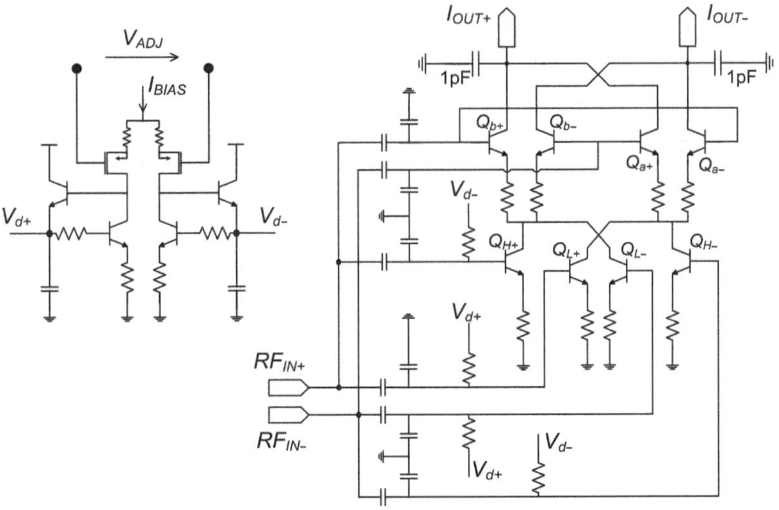

Fig. 2.48 Adjustable power detector topology

(Q_{H+}, Q_{H-}) and (Q_{L+}, Q_{L-}) in Fig. 2.48. Compared to other detector topologies such as logarithmic detectors, mixer-based detectors present a relatively low dynamic range (<50 dB). However, in the present application, this is considered as sufficient. A positive value for the adjustment voltage V_{ADJ} allows preferably biasing the pair

(Q_{H+}, Q_{H-}) and the mixer conversion gain is positive ; whereas a negative value for the adjustment voltage V_{ADJ} allows preferably biasing the pair (Q_{L+}, Q_{L-}) and the mixer conversion gain becomes negative.

When $V_{ADJ} = 0$, both pairs are equally biased and mutually thwart. The conversion gain is close to nil in this case. It is therefore possible to gradually control the mixer conversion gain over a wide range. From a theoretical point of view, the output differential current can be derived (identically to what is done with a classic Gilbert cell) as follows:

$$I_{MIX} = I_{BIAS} \cdot \frac{\sinh^2\left(\frac{V_{IN} \cdot A_{att}}{2U_t}\right)}{\cosh\left(\frac{V_{IN} \cdot A_{att}}{2U_t}\right)} \cdot \tanh\left(\frac{V_{ADJ}}{2U_t}\right) \tag{2.35}$$

where V_{IN} is the RF input voltage,
V_{ADJ} is the mixer conversion gain adjustment voltage,
I_{BIAS} is the mixer bias current,
A_{att} is the attenuation of the input capacitive divider
U_t is the thermal voltage

Equation (2.35) can be approximated at low power levels and low V_{ADJ} voltage by:

$$I_{MIX} = I_{BIAS} \cdot \frac{A_{att}^2 \cdot V_{IN}^2 \cdot V_{ADJ}}{8U_t^3} \tag{2.36}$$

In order to make the output detector as little dependent of temperature as possible, the current I_{BIAS} is generated by a *Proportionnal-to-Ambient-Temperature* source.

The simulated response of the differential output current vs. input *RF* magnitude is represented on the following logarithmic scale graph (Fig. 2.49) for various V_{ADJ} values and with a *CW RF* input signal. The linear shift demonstrates that the output current is linearly controlled by the adjustment voltage. It must be remarked that at $V_{ADJ} = 0$ V, the differential output current is not nil which is caused by the mixer *DC* offset.

2.5.4 *Management of Linearity/Efficiency Trade-Off via 3-Bit Delta-Sigma-Like Closed-Loop Reconfigurable PA*

As *WLAN* application is still considered, clock frequency is chosen as high as 160 MHz to have $OSR = 4$. Figure 2.50 depicts the simulated reconfigurable closed-loop response in frequency domain (no output filter included) with 1-bit and 3-bit comparators at a 15 dBm output power and applying a *WLAN* modulated signal.

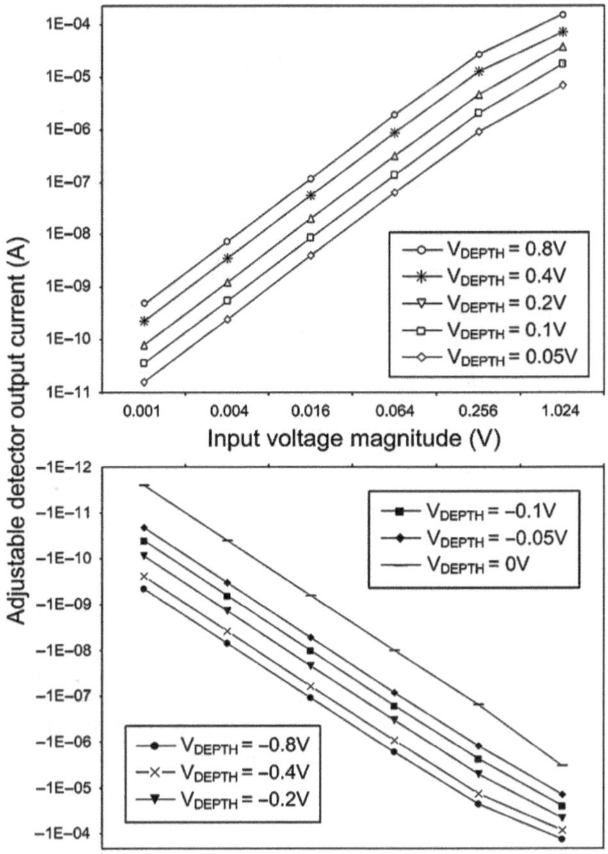

Fig. 2.49 Power detector response for various V_{ADJ} values

The dotted line stresses the wide-band noise shaping. The threshold voltage non-linear distribution of the 3-bit comparator is optimized from the stand-point of the efficiency/linearity tradeoff by setting the exponent to $\gamma = 2$.

This illustrates the important effect of the resolution on the noise/distortion level. Here *ACLR* is decreased from –27 to –32 dBc, i.e. a 5 dBc improvement per a 2bit increase in resolution. This means that the classic law (2.37) that expresses the *Signal-to-Noise/Distortion-Ratio* (*SNDR*) as a function of the resolution in $\Delta\Sigma$ *ADC* can no longer be applied for the theoretical calculus of *ACLR* in such reconfigurable system:

$$SNDR = 6.02\,M + 1.76 \tag{2.37}$$

where M is the resolution.

As expected in Section 2.5.2.3, over the loop operation power range (3–25 dBm output power), *ACLR* is system-controlled and output power independent. It is

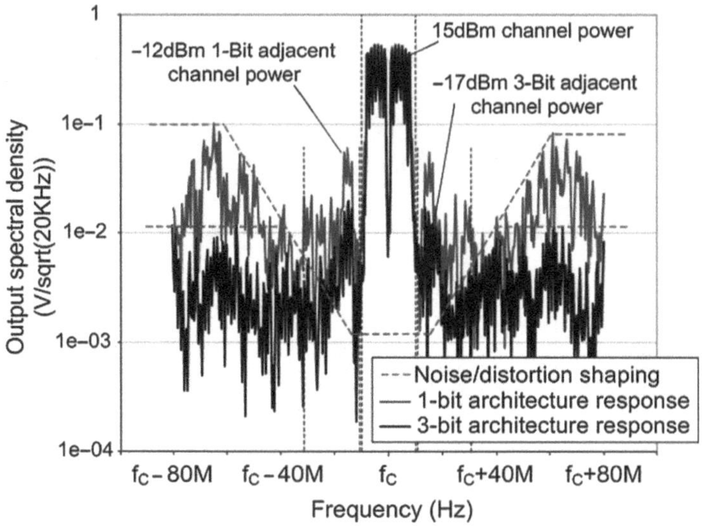

Fig. 2.50 Output *RF* spectrum (no filter) with 1-bit (*clear line*) and 3-bit (*dark line*) architectures (@P_{OUT} = 15 dBm)

forced to a ~30 dBc value (Fig. 2.51, dark line). This means that the loop properly works. The average *PAE* is significantly improved compared to the full-power case (8 active cells) at medium power levels. The maximum *PAE* is found to be 30% (not depicted here).

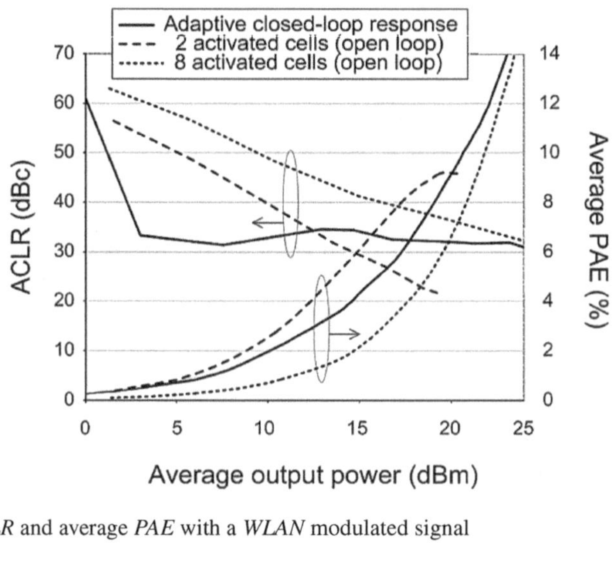

Fig. 2.51 *ACLR* and average *PAE* with a *WLAN* modulated signal

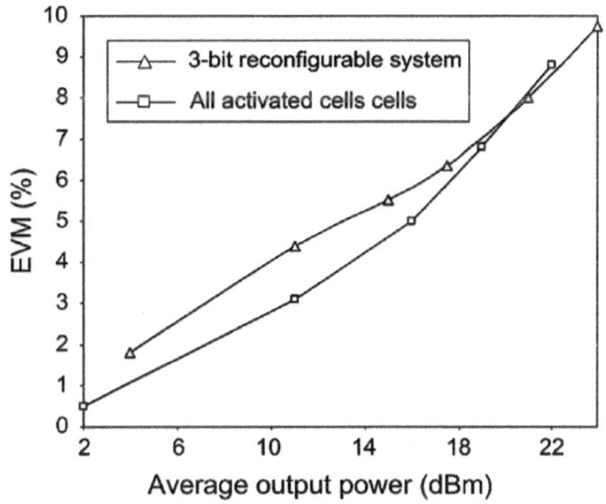

Fig. 2.52 Average *EVM* with a *WLAN* modulated signal

The waveform *Error Vector Magnitude vs. Output power* is depicted in Fig. 2.52. *EVM* with a 3-bit reconfigurable architecture grows monotonously because phase distortion is not handled by the loop.

However a slight inflection is observed at medium power levels which illustrates the closed-loop effect. The *EVM* with a 3-bit reconfiguration closed-loop system presents moderate degradation (1.3% max.) compared with a full power amplifier (when all eight cells are active). Above all, the increase in *EVM* at low power levels that was observed with an open loop system (see Section 2.4.3.5) is corrected.

Table 2.1 Comparative performance/features of the architectures developed in Sections 2.4 and 2.5

	ΔΣ-*BICS*-controlled *PA*	ΔΣ-like closed-loop *PA*
Linearity	Optimum performance in the middle of reconfiguration power range	Better control over power range due to closed-loop
PAE	Comparable	
Sensitivity to *VSWR*	Moderate sensitivity and increased *PA* robustness (see Section 2.4.3.6) +	Very sensitive → Second loop necessary (see Section 2.5.3.1)
Non-linear distribution technique	Non-linear *DAC* feedback	Non-linearly distributed threshold voltages
Digital decimation filter (prior to power core)	Possible under latency constraints +	Impossible
Calibration	Yes I_{FLOOR}, I_{DAC}	Not necessary +

2.5.5 *Conclusion and Comparison with Delta–Sigma BICS-Controlled Architecture*

The approaches that were developed in Sections 2.4 and 2.5 present various advantages and weaknesses that are summarized in Table 2.1.

2.5.6 *Prospect Works Based on These Techniques*

2.5.6.1 Combined Cartesian/Reconfigurable Dual Loop Architecture

The main issue in the architectures that were presented in Sections 2.4 and 2.5 is that they are based on envelope detection. Phase distortion is therefore not taken into account and corrected. In order to tackle this issue, combining Cartesian Feedback

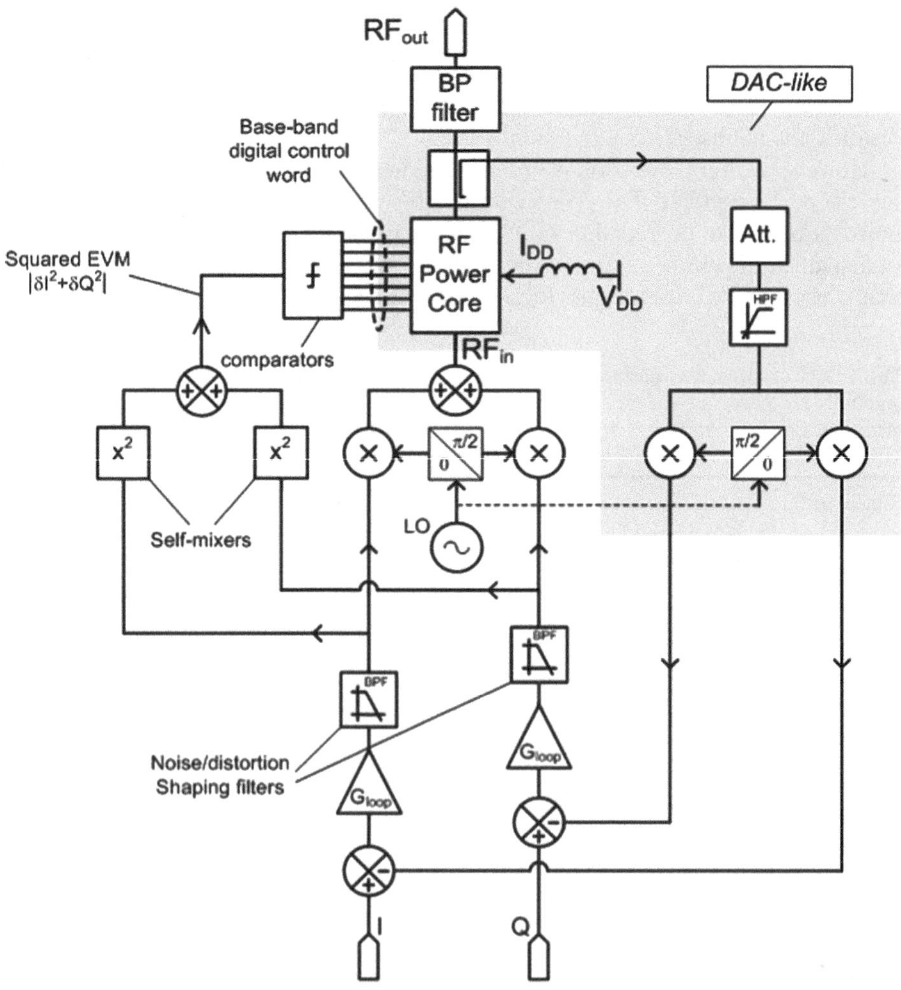

Fig. 2.53 Cartesian/reconfigurable dual-loop architecture

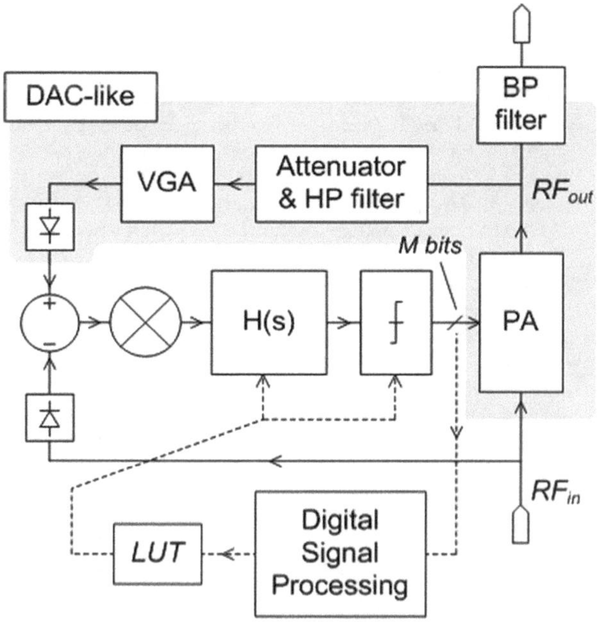

Fig. 2.54 Linearity monitoring and example of system reconfiguration by means of a Look-Up-Table (*LUT*)

principle with the reconfigurable system that was proposed in Section 2.5 is a promising prospect. The synoptic of this approach is depicted in Fig. 2.53. One of the forward paths acts on the *I/Q* signal itself and plays the role of linearization only. The second one reconfigures the power core, plays the role of adaptive efficiency enhancement, and indirectly impacts on the *RF* output.

2.5.6.2 Linearity Monitoring via Digital Signal Processing

The advantage of the topologies in Sections 2.4 and 2.5 lies in the fact that a digital control word is used. When desired, this data can be treated by an external *Digital Signal Processing* as depicted in Fig. 2.54. This is at the cost of higher complexity, but this way, it becomes possible to calculate and dynamically monitor the in-band or out-of band spurious levels, and to reconfigure/adjust some system parameters as a function of the targeted application or data rate/modulation scheme/constellation.

Indeed, over-sampling ratio, I_{FLOOR}, I_{DAC} in the example of the *BICS* detection, or the threshold voltages in the example described in Section 2.5 can be dynamically reconfigured when an excessive linearity degradation is monitored.

References

Cherry JA (1994) Distortion analysis of weakly nonlinear filters using. Volterra series. M.Eng. thesis, Carleton University, Ottawa, ON, Canada

Deng J, Gudem PS, Larson LE, Asbeck PM (2005) A high average-efficiency SiGe HBT power amplifier for WCDMA handset applications. Trans Microw Theory Tech 53(2):529–537. doi:10.1109/RFIC.2007.380826

Huang PC, Chen YH, Wang CK (2000) A 2-V 10.7-MHz CMOS limiting amplifier/RSSI. J Solid-State Circuits 35(10):1474–1480. doi:10.1109/4.871325

Khoo KBH (1998) Programmable high-dynamic range sigma–delta A/D converters for multistandard, fully-integrated RF receivers. MEng thesis, UC Berkeley, CA, USA

Leyssenne L, Jarry P, Pham JM, Kerhervé E, Saias D (2005a) Design of a multi-cell power amplifier for WCDMA application. Proc Int Microw Optoelectron Conf, Brasilia, Brazil. pp 112–116. doi:10.1109/IMOC.2005.1580094

Leyssenne L, Jarry P, Pham JM, Kerhervé E, Saias D, Vigne A (2005b) Design of a differential WCDMA power amplifier. In: Jarry P (ed) Microwave filters and amplifiers. Research Signpost, Kerala. ISBN:81-308-0009-8

Leyssenne L, Kerhervé E, Deval Y, Belot D (2008) A SiGe power amplifier dedicated to power management for 802.11n/802.16e Standard. Top Workshop Power Amplif Wirel Commun, Orlando, USA

Leyssenne L, Kerhervé E, Deval Y, Belot D (2009a) A novel WLAN power amplifier adaptive loop based on delta–sigma non-linearity control. Proc Radio Week Symp, San Diego, USA. pp 594–597. doi:10.1109/RWS.2009.4957421

Leyssenne L, Kerhervé E, Deval Y, Belot D (2009b) A novel delta sigma built-in-current-sensor as a signal strength indicator for RF transceiver reconfiguration. Proc Symp Int Circuits Syst Des, Natal, Brazil. pp 69–73

Maidon Y, Deval Y, Begueret JB (2000) An improved CMOS BICS for on-line testing. Proc Int On-Line Test Workshop, Palma de Mallorca, Spain. pp 100–103. doi:10.1109/OLT.2000.856620

Presti D, Carrara F, Palmisano G, Scuderi A (2008) A high-resolution 24-dBm digitally-controlled CMOS power amplifier for multi-standard RF polar transmitters. Proc Eur Solid-State Circuits Conf, Edinburgh, UK. pp 482–485. doi:10.1109/ESSCIRC.2008.4681897

Su DK, McFarland WJ (1998) An IC for linearizing RF power amplifiers using envelope elimination and restoration. J Solid-State Circuits 33(12):2252–2258. doi:10.1109/4.735710

Vázquez JR, Pineda de Gyvez J (2004) Built-in current sensor for IDDQ testing. J Solid-State Circuits 39(3):53–58. doi:10.1109/VTEST.2004.1299225

Zorn S, Ehm HJ, Weigel R (2008) A novel technique for determining kernels of Volterra based behavioral models for RF amplifiers. Eur Microw Conf, Amsterdam, The Netherlands. pp 246–249. doi: 10.1109/EUMC.2008.4751434

Chapter 3
Continuous Adaptive Bias Technique for Radiofrequency Power Amplifiers

Abstract Chapter 3 proposes an adaptive-bias power amplifier that is based on a finely tuned envelope injection principle. Such flexible technique makes a simultaneous improvement of linearity and efficiency possible at the cost of minor complexity, reduced die area, and no extra bill-of-material. To this end, a control voltage V_{DEPTH} must be calibrated (e.g., via a look-up table) as a function of the average power level and of the power device characteristics. Though not technology-restricted, this technique is validated via a differential *LDMOS PA* module whose output load network is built with a 4th order band-pass *Integrated Passive Device*. First of all, the general context of this work is presented with the various *TX* circuits that are associated with the power amplifier. Then, the theory of the adaptive bias mechanism is explained in time and frequency domains by means of the *non-linear-current method*. Memory effects are also theoretically evoked and their impact on adaptive bias power amplifiers is illustrated. The advantages and the limitations of the proposed topology in terms of linearity/efficiency enhancement are highlighted by various co-simulations and characterizations (*on-board* and *on-wafer*) with *EDGE* and *WCDMA* modulated signals.

Key words *PA* module assembly · Envelope Tracking · Adaptive bias · Fine tuning of envelope injection · Analog linearization · Memory effects · Spectral regrowth unbalance · Integrated passive device design · *DCS/EDGE/WCDMA* demonstrator on silicon · Load-pull characterization

3.1 Introduction and Theory

3.1.1 Introduction

The *PA* architecture that is presented here was proposed in the frame of *FP6 STREP IST MOBILIS* European project (MOBILIS 2006). It consists in a dual-band *DCS/EDGE/WCDMA* transceiver (Fig. 3.1) which utilizes an *I/Q* modulator associated with *BAW* filters (Flament et al. 2009), a wide-band adaptive power amplifier, a *BAW* duplexer (Shirakawa et al. 2006), and *SOI* switches.

L. Leyssenne et al., *Reconfigurable RF Power Amplifiers on Silicon for Wireless Handsets*, Analog Circuits and Signal Processing, DOI 10.1007/978-94-007-0425-1_3,

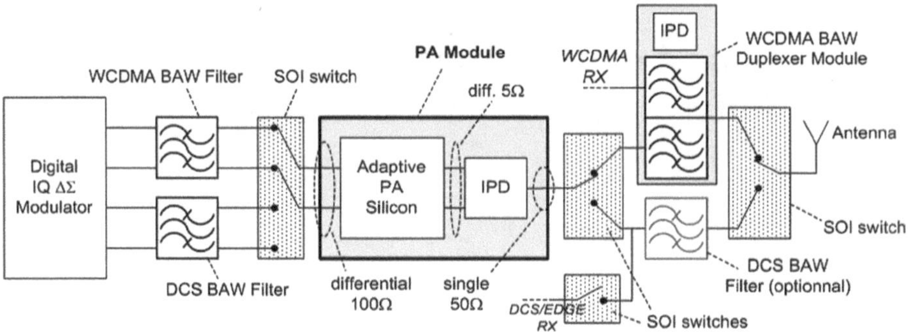

Fig. 3.1 *MOBILIS* project synoptic

The dual-band *I/Q* modulator is built on a novel differential digital ΔΣ architecture that maximizes the bandwidth and dynamic range at the cost of having low consumption. It was developed with *CMOS* 90 nm ST Microelectronics technology. Band-pass filters are inserted at the modulator output in order to loosen the specifications in terms of duplexer out-of-band rejection and *PA* linearity. For sake of optimization, the *I/Q* modulator and the *BAW* filters (resp. the duplexer module) were preferably developed according to a differential topology (resp. a single-ended topology). The duplexer module consists in an *Integrated Passive Device* (*IPD*) over which *TX* and *RX* band-pass *BAW* filters are flip-chipped. *BAW* and *IPD* technologies were provided by CEA–LETI and ST Microelectronics respectively.

The Power Module is at the crossroads of all these blocks and its design must take into account various constraints. Above all, its power capability must be high enough to overcome the power losses of approximately 3 dB in the duplexer and the *SOI* switches. For 2nd and 3rd generation standards, the *Peak-to-Average Power Ratio* is a relevant estimation of the power back-off that must be taken into account to determine the *PA* continuous-wave performance as shown in (3.1).

$$\begin{aligned} CP_{1dB} &\approx P_{avg\_antenna} + \underbrace{back - off}_{\approx PAPR} + IL_{duplexer+switches} \\ &\approx \begin{cases} 26\,\text{dBm} + 3\,\text{dB} + 3\,\text{dB} = 32\,\text{dBm}\,\textit{for EDGE} \\ 24\,\text{dBm} + 3\,\text{dB} + 3\,\text{dB} = 30\,\text{dBm}\,\textit{for WCDMA} \end{cases} \end{aligned} \tag{3.1}$$

The *PA* module consists in the wire-bonded assembly of a differential adaptive-bias silicon die and a specific *IPD* (Fig. 3.2).

This *IPD* plays the role of an impedance matching network and carries out the differential-to-single-ended conversion. Though such assembly is compliant with *SIP* technology, only *Chip-On-Board* (*COB*) will be considered throughout this chapter.

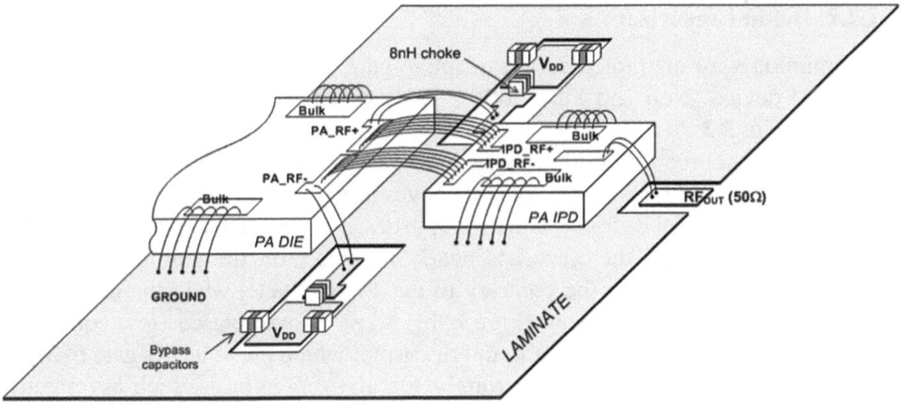

Fig. 3.2 *PA* module assembly

3.1.2 Adaptive Power Amplifier Principle and Architecture

3.1.2.1 Principle

The handset power amplifier architecture that is presented here is based on the *Continuous Envelope Tracking* principle by means of dynamic gate bias. Several works have successfully explored this technique (Cha et al. 2003, Sahu 2004, Kim et al. 2004, Noh and Park 2004, Deng 2005, Tanoue et al. 2005, Oka et al. 2005). Dynamic gate bias is designed to maximize the average efficiency of 3G handset power amplifiers at the expense of low complexity and reduced silicon area. Indeed, *Envelope Elimination* and *Restoration* or *Drain Bias Envelope Tracking* are the most commonly employed and provide the most enhanced efficiency response to date (see Sections 1.2.3.1 and 1.2.3.2). These techniques imply the use of a large *DC/DC* converter and at least one external choke inductor that are not necessarily compliant with the low area and low *Bill-of-Material* requirements of handset manufacturers. Moreover, the magnitude/phase synchronization condition is one more difficulty that must be overcome when *EER* technique is chosen. As opposed to the architectures based on fragmented power amplifiers in Chapter 2, the adaptive bias principle that is explored in this section consists in continuously and dynamically shifting the *PA* bias current level according to power variations. As *WCDMA* and *EDGE* are more specifically targeted, this adaptive bias mechanism must be able to handle fast envelope swing (3 dB peak-to-average ratio at 3.84 Mcps for *WCDMA*) and broad average power dynamic range (up to 70 dB for *WCDMA*). In order to save-battery life-time, the quiescent bias level (i.e. when no *RF* power is applied to the *PA*) must be as low as possible. If the quiescent bias level is too low, the power stage presents poor performances in terms of noise and linearity. Therefore, strong power gains modulation in either magnitude (*AM/AM*) or in phase (*AM/PM*) is observed. The goal of the proposed architecture is precisely to improve efficiency at low power levels while minimizing linearity degradation.

3.1.2.2 Diode-Linearizer

One common way of implementing adaptive gate bias consists in using a diode-connected device (Noh and Park 2004). This approach is used with the *LDMOS* devices in Fig. 3.3.

A common-source *LDMOS* M_{LDa} is used as a power device. Its gate is biased by M_{LDb}, a common-gate *LDMOS* that behaves as a non-linear load and carries out predistortion. When the driver transistor pushes a high current peak (Fig. 3.3a), M_{LDb} switches off and therefore has nearly no effect on the gate charge of the power transistor M_{LDa}. On the contrary to the charge cycle, when the driver transistor draws a high current peak (Fig. 3.3b), M_{LDb} is over-biased (low equivalent impedance) and provides a large amount of current which prevents the gate of M_{LDa} device from being discharged too strongly. On an average basis, such asymmetric charge/discharge cycle results in an increase of M_{LDa} gate voltage and consequently an increase of the average bias current I_{DD} which enhances the *PA* power capability. One of the useful properties of such detection law is that it makes the power stage intrinsically more robust to output load variations. If a sudden positive step in voltage gain (and therefore Miller capacitance) is encountered due to a *VSWR* variation, a drop in the *RF* gate voltage amplitude and consequently in the average bias current of the power transistor is observed. The *PA* is thus protected from high output voltage swings. Such analog predistortion is all the more efficient as

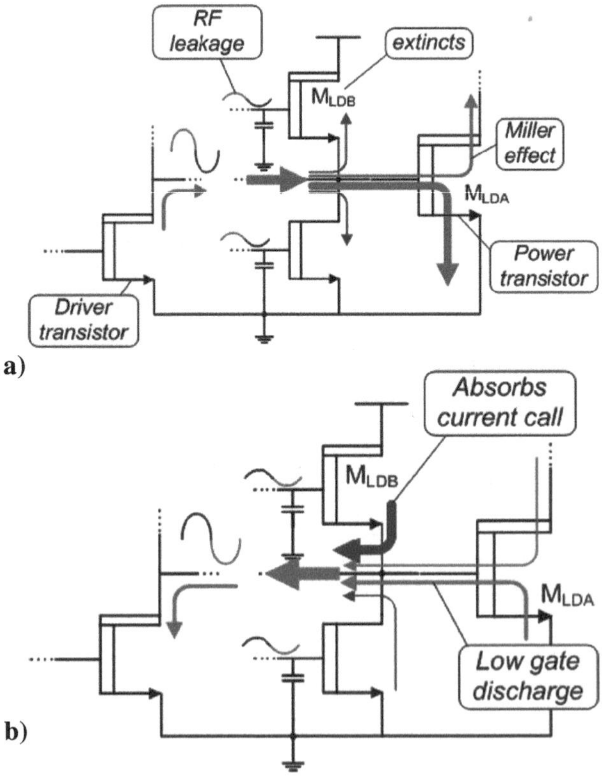

Fig. 3.3 Illustration of (a) charge and (b) discharge cycles of a diode-linearized power stage

the *diode-linearizer* is matched with the power device. In the following sections, both M_{LDa} and M_{LDb} are implemented with ST Microelectronics 2nd generation *LDMOS*.

In order to gain a better insight on the way adaptive bias operates, a practical mathematical formalism in frequency domain based on the *non-linear current method* is often used (Maas 2003). This consists in treating a non-linear problem via a linear small-signal analysis and superposing some additional non-linear components that are derived by Taylor developments. The small-signal expression of M_{LDa} input gate admittance is inferred from Fig. 3.3 and (B.3) in the theoretical analysis of Appendix B (index a refers to M_{LDa} device):

$$Y_{g,a}(\omega) = j\omega \cdot \left(C_{gs,a} + C_{gd,a} \cdot (1 + A_V(\omega))\right) \tag{3.2}$$

where A_v is the voltage gain of M_{LDa}.

Likewise, the small-signal expression of M_{LDb} input admittance is (index b refers to M_{LDb} device):

$$Y_{g,b}(\omega) = \frac{g_{m,b} + j\,C_{gs,b}\,\omega}{1 + j\,r_g\,C_{gs,b}\,\omega} \tag{3.3}$$

In the following paragraphs, the terms $Y_{g,a}$ and $Y_{g,b}$ will be considered as non-linear admittance expressions in frequency domain. Their Volterra kernels $Y_{g,a,j}$ and $Y_{g,b,j}$ will be developed by proper derivation from their respective linear (1st order) expressions (3.2) and (3.3). For sake of simplicity, the analysis will be limited to 3rd order and the gate input signal will be considered as a dual-tone voltage whose magnitude, carrier frequency and frequency spacing are respectfully noted V_g, ω_C and $\Delta\omega$, where $\Delta\omega << \omega_C$. Throughout this chapter, the numbers used as indices represent the order of non-linearity. By substituting the terms $C_{gs,a}$, $C_{gd,a}$, and A_V with their non linear counterparts, the Volterra kernels of $Y_{g,a}$ are obtained:

$$Y_{g,a,1}(\omega_C) = j\omega_C \cdot \left(C_{gs,1} + \underbrace{C_{gd,1} \cdot \left(1 + A_{V,1}(\omega_C)\right)}_{\textit{Miller effect}} \right) \tag{3.4}$$

$$Y_{g,a,2}(\omega_C, \Delta\omega - \omega_C) = j\,\Delta\omega \begin{pmatrix} C_{gs,a,2} + C_{gd,a,2} \cdot \left(1 + A_{V,1}(\omega_C)\right) \\ + C_{gd,a,1}\,A_{V,2}(\omega_C, \Delta\omega - \omega_C) \end{pmatrix} \tag{3.5}$$

$$Y_{g,a,3}(\omega_C, \omega_C, \Delta\omega - \omega_C) \approx j\omega_C \begin{pmatrix} C_{gs,3} + C_{gd,3} \cdot \left(1 + A_{V,1}(\omega_C)\right) \\ + C_{gd,2} \cdot A_{V,2}(\omega_C, \Delta\omega - \omega_C) \\ + C_{gd,1}\,A_{V,3}(\omega_C, \omega_C, \Delta\omega - \omega_C) \end{pmatrix} \tag{3.6}$$

By substituting the terms $C_{gs,b}$, $C_{gd,b}$, and $g_{m,b}$ with their non linear counterparts, the Volterra kernels of $Y_{g,b}$ are obtained:

$$Y_{g,b,1}(\omega_C) = \frac{g_{m,b,1} + r_g\,\omega_C^2\,C_{gs,b,1}^2 + j\,C_{gs,b,1}\,\omega_C \times \left(1 - g_{m,b,1}\,r_g\right)}{\left(1 + \omega_C^2\,r_g^2\,C_{gs,b,1}^2\right)} \tag{3.7}$$

$$Y_{g,b,2}(\omega_C, \Delta\omega - \omega_C) = -\frac{\left(\begin{array}{l}(g_{m,b,2} + 2\, C_{gs,b,2}\, C_{gs,b,1}\, \omega_C\, \Delta\omega) \\ \quad + j\Delta\omega \cdot \left(C_{gs,b,2} + r_g \times \left(g_{m,b,2}\, C_{gs,1} + g_{m,b,1}\, C_{gs,b,2}\right)\right)\end{array}\right)}{\left(1 + \omega_C^2\, r_g^2\, C_{gs,b,1}^2\right)} \tag{3.8}$$

$$Y_{g,b,3}(\omega_C, \omega_C, \Delta\omega - \omega_C) = \frac{\left(\begin{array}{l}g_{m,b,3} - \omega_C\, r_g \left(\omega_C\, C_{gs,b,1}\, C_{gs,b,3} + \Delta\omega \cdot C_{gs,b,2}^2\right) \\ \quad + j\left(\omega_C \cdot C_{gs,b,3} \cdot \left(1 - g_{m,b,1} \cdot r_g\right) - g_{m,b,2} \cdot r_g \cdot \Delta\omega \cdot C_{gs,b,2}\right)\end{array}\right)}{\left(1 + \omega_C^2\, r_g^2\, C_{gs,b,1}'^2\right)} \tag{3.9}$$

From the stand-point of M_{LDa} gate, the schematic in Fig. 3.3 can be modeled by the equivalent circuit in Fig. 3.4, assuming that the input gate voltage magnitude in *RF* domain is V_{g0}. The 2nd order non-linear components determine the *adaptive bias mechanism*, i.e. the gate voltage magnitude in the base-band domain, namely at the frequency $\Delta\omega$ (3.10).

$$V_g(\Delta\omega) = \alpha(\Delta\omega) \times \frac{\left|V_{g0}\right|^2}{2} \tag{3.10}$$

where

$$\alpha(\Delta\omega) = -\frac{\left(Y_{g,a,2}(\omega_C, \Delta\omega - \omega_C) + Y_{g,b,2}(\omega_C, \Delta\omega - \omega_C)\right)}{\left(Y_{g,a,1}(\Delta\omega) + Y_{g,b,1}(\Delta\omega)\right)} \tag{3.11}$$

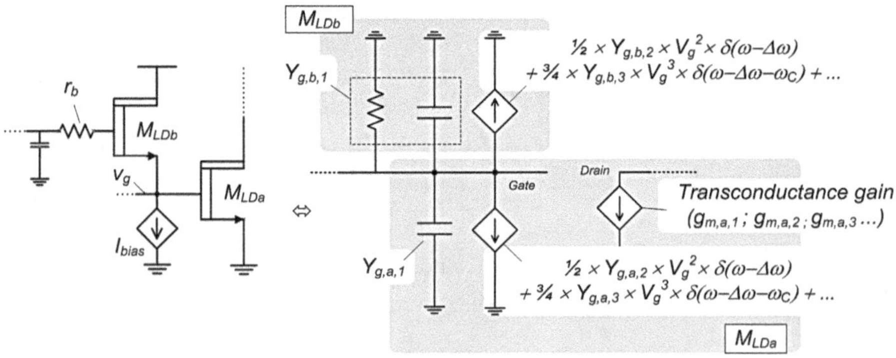

Fig. 3.4 Diode linearization modelling via 2nd and 3rd order non-linear current sources

The term $\alpha(\Delta\omega)$ is the *basic power adaptability factor*. In this context 'basic' means that it is intrinsic to the power device and its bias transistor and is independent of any external system. It expresses the relationship between the *RF* envelope and the low frequency gate voltage modulation. By substituting (3.4), (3.5), (3.7), and (3.8) in (3.11) and (3.10), the average gate voltage deviation due to adaptive bias is obtained (3.12) and is found to be positive and proportional to power. In Chapter 1, it was stated that *LDMOS* devices feature a positive 2nd order transconductance gain g_{m2} over a wide bias range.

$$\Delta V_g(0) = \frac{g_{m,b,2}}{g_{m,b,1} + r_g\,\omega_C^2\,C_{gs,b,1}^2} \times \frac{\left|V_{g0}\right|^2}{2} \approx \frac{g_{m,b,2}}{g_{m,b,1}} \times \frac{\left|V_{g0}\right|^2}{2} > 0 \tag{3.12}$$

The term $\alpha(\Delta\omega)$ has a frequency-dependent behavior. Hence, for high spacing bandwidth $\Delta\omega$, the modulus of $\alpha(\Delta\omega)$ drops and a phase shift between the base-band gate voltage and the actual *RF* envelope arises. Fortunately, the signs of the 2nd order imaginary components $\Im(Y_{g,a,2})$ and $\Im(Y_{g,b,2})$ are opposite one to the other. Indeed, as *LDMOS* devices present positive 2nd degree capacitances and transconductance, it can be inferred from (3.5) and (3.7) that:

$$\Im\left(Y_{g,a,2}(\omega_C, \Delta\omega - \omega_C)\right) = \Delta\omega \begin{pmatrix} C_{gs,a,2} + C_{gd,a,2}\cdot\left(1 + A_{V,1}(\omega_C)\right) \\ + C_{gd,a,1}\cdot\Re\left(A_{V,2}(\omega_C, \Delta\omega - \omega_C)\right) \end{pmatrix} > 0$$

And that

$$\Im\left(Y_{g,b,2}(\omega_C, \Delta\omega - \omega_C)\right) = -\frac{\Delta\omega\cdot\left(C_{gs,b,2} + r_g \times \begin{pmatrix} g_{m,b,2}\,C_{gs,1} \\ + g_{m,b,1}\,C_{gs,b,2} \end{pmatrix}\right)}{\left(1 + \omega_C^2\,r_g^2\,C_{gs,b,1}^2\right)} < 0$$

This means that the *diode linearizer* plays a role of synchronization. By properly biasing the device M_{LDb} and its ballast gate resistor r_g, it becomes possible to minimize the phase deviation in $\alpha(\Delta\omega)$ over a wide bandwidth (at least 5 MHz in the case of *WCDMA*).

The impact of the *diode linearizer* also intervenes in the analysis of 3rd degree non-linearities. Indeed, a 3rd order inter-modulation voltage appears at the gate and at frequency $(\omega_C + \Delta\omega)$ and is given by expression (3.13):

$$V_g(\omega_C + \Delta\omega) = \kappa(\omega_C, \omega_C, \Delta\omega - \omega_C) \times \frac{3}{4} \times V_{g0}^2 \cdot V_{g0}^* \tag{3.13}$$

where

$$\kappa(\omega_C, \omega_C, \Delta\omega - \omega_C) = -\left(\frac{Y_{g,a,3}(\omega_C, \omega_C, \Delta\omega - \omega_C) + Y_{g,b,3}(\omega_C, \omega_C, \Delta\omega - \omega_C)}{Y_{g,a,1}(\omega_C) + Y_{g,b,1}(\omega_C)}\right) \tag{3.14}$$

As was stated of 2nd order components, the signs of the 3rd order imaginary components $\Im(Y_{g,a,3})$ and $\Im(Y_{g,b,3})$ are opposite one to the other. As *LDMOS* devices present negative 3rd degree capacitances and transconductance, it can be inferred from (3.6) and (3.8) that

$$\Im\left(Y_{g,a,3}\left(\omega_C,\omega_C,-\omega_C\right)\right)\approx\omega_C\left(\begin{array}{l}C_{gs,3}+C_{gd,3}\cdot\left(1+A_{V,1}\left(\omega_{RF}\right)\right)\\ \qquad +C_{gd,1}\,A_{V,3}\left(\omega_C,\omega_C,-\omega_C\right)\end{array}\right)<0$$

and

$$\Im\left(Y_{g,b,3}\left(\omega_C,\omega_C,-\omega_C\right)\right)=\frac{\omega_C\cdot C_{gs,b,3}\cdot\left(1-g_{m,b,1}\cdot r_g\right)}{\left(1+\omega_C^2\,r_g^2\,C_{gs,b,1}^{\prime 2}\right)}>0$$

The *diode linearizer* allows the imaginary part of $\kappa(\omega_C,\omega_C,-\omega_C)$ to be decreased/cancelled whereby the *CW* phase modulation (*AM/PM*) of the power amplifier at frequency ω_C is reduced. As it will be demonstrated in Section 3.3.2.3, diode linearization introduces an inflection point in the *AM/PM* response. The real part of $\kappa(\omega_C,\omega_C,-\omega_C)$ has a positive sign (since $-g_{m,a,3} > 0$) and plays a marginal role in the compensation of the 3rd order transconductance gain $g_{m,a,3}$ (see Section 3.1.2.4).

3.1.2.3 Overall *PA* Adaptive Bias Architecture

The main goal of the proposed dynamic gate bias technique is that it enables simultaneous optimization of linearity and efficiency. In practice, such a condition is difficult to fulfil both over wide channel bandwidths and power ranges. To this end, the envelope injection must be accurately controlled in magnitude and phase over the whole complex plan. Several injection schemes have been proposed based on the mutual compensation of non-linear components for example 2nd and 3rd order components, or 3rd and 5th components (Wang 2003, Paulin et al. 2004, Mizusawa and Kusunoki 2005, Deng 2005, Leung et al. 2005). In this way, all non linear components can be destructively combined thus improving linearity at the cost of minimal bias current increase. Those techniques provide good results but are optimized in the vicinity of a specific power level. In order to enhance the linearity response over a wider power range, it is necessary to make the envelope injection adjustable. To this end, a dual-adaptive injection system was proposed in the form of a patent (Leyssenne et al. 2007) and is depicted in Fig. 3.5. In this book, the proposed topology is applied to a *LDMOS* adaptive differential power amplifier (0.25 μm *BICMOS* ST Microelectronics technology) dedicated to *DCS/EDGE/WCDMA* applications even though this architecture is not exclusive in terms of technology/device/standard. It is composed of a *coarse bias circuit* and a *fine tuning bias system*.

The *coarse bias circuit* is implemented by means of the diode-connected *LDMOS* M_{LDb} that was previously described in Section 3.1.2.2. This coarse bias system is

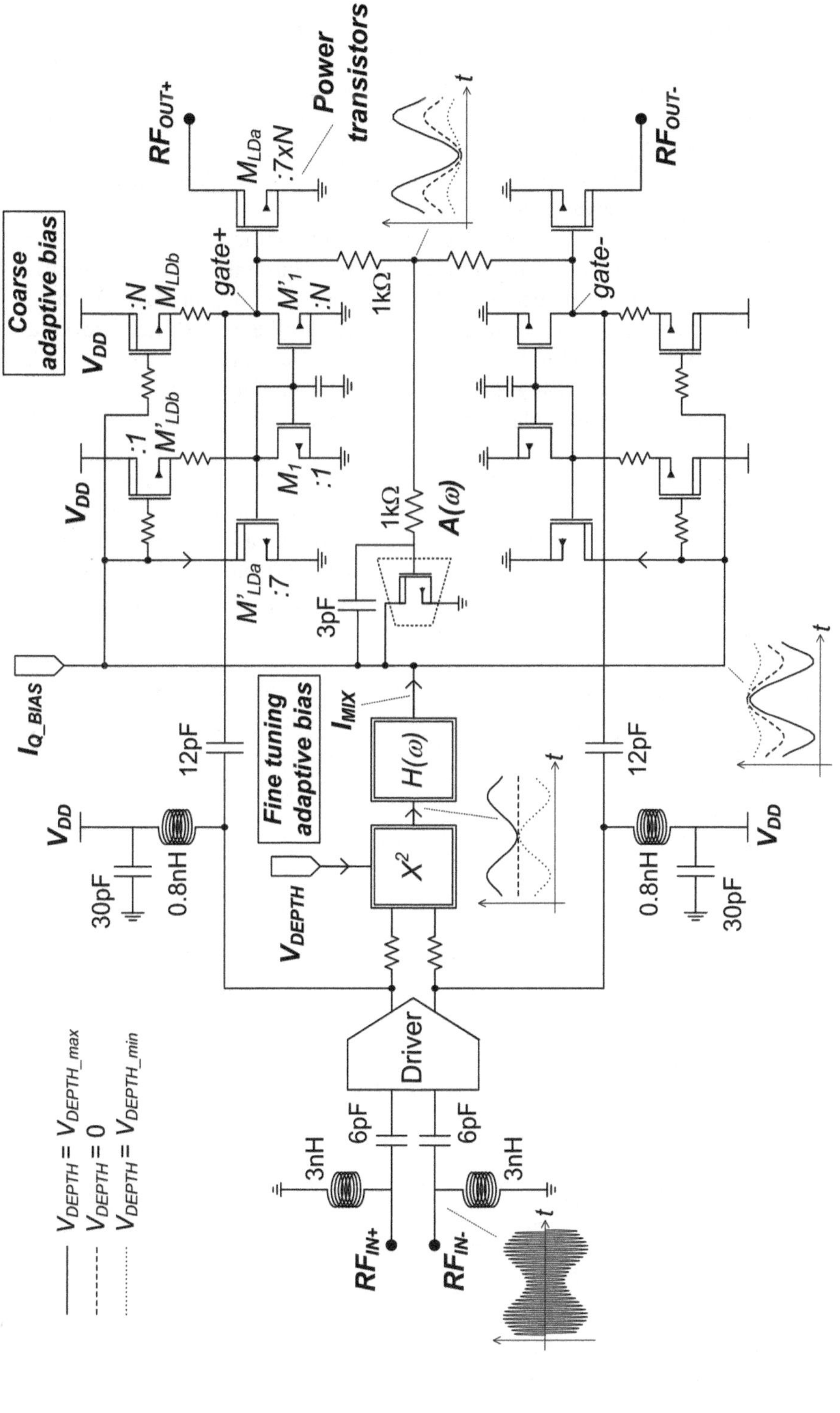

Fig. 3.5 Dual adaptive bias power stage (Leyssenne et al. 2007)

frequency-limited as it was stated from (3.11), in particular due to the 2nd-order non-linear capacitances $C_{gs,2}$ of M_{LDa} and M_{LDb}. In order to enhance the coarse envelope tracking bandwidth (that would otherwise be in the order of a few MHz), a closed-loop bias system is required. This was carried out by inserting a feedback transconductance $A(\Delta\omega)$ that introduces a zero in the closed-loop transfer function. This feedback network behaves as a Common-Mode Feedback and also provides robustness to temperature variations.

The *finely tuned bias system* is designed to supply/withdraw a gradually controlled (in magnitude and phase) additional envelope current in order to improve linearity. It is built with an adjustable self-mixer (see X^2 block in Fig. 3.5) that is controlled by the voltage V_{DEPTH}, and an adjustable base-band filter (see $H(\omega)$ filter in Fig. 3.5) that is controlled by the voltage V_{Φ}. The adjustable self-mixer is implemented according to the topology in Section 2.5.3.2. The only difference is that its differential output current is folded into a single-ended output prior to driving the base-band filter.

The magnitude and polarity ($\pm 180°$) of the envelope current that is injected by the self-mixer are determined by V_{DEPTH} that can be set either as positive or negative if envelope reversal is desired. In Fig. 3.5, simple dual-tone chronograms are depicted at various nodes to illustrate to what extent the depth of adaptive bias can be adjusted. Three cases are considered: the extrema $V_{DEPTH} = V_{DEPTH\_min} < 0$, and $V_{DEPTH} = V_{DEPTH\_max} > 0$ as well as the mid-point $V_{DEPTH} = 0$ when *fine tuning* is not active.

The proposed adjustable filter $H(\omega)$ is built on the topology in Fig. 3.6 and is compliant with *CMOS* technology. This filter operates in current-mode and is a closed-loop circuit where the block $B(s)$ behaves as an integrator whose time constant is τ_0.

The synoptic of this filter is given by Fig. 3.7. Its frequency response can be gradually reconfigured by means of an adjustment voltage V_{Φ}.

$$H(\omega)|_{V_\phi} = \left(\frac{I_{OUT}}{I_{IN}}\right)(\omega)\Bigg|_{V_\phi} = h_0^\circ(V_\phi) \times \frac{\left(1 + \frac{j\omega}{\omega_Z(V_\phi)}\right)}{\left(1 + \frac{j\omega}{\omega_p}\right)} \tag{3.15}$$

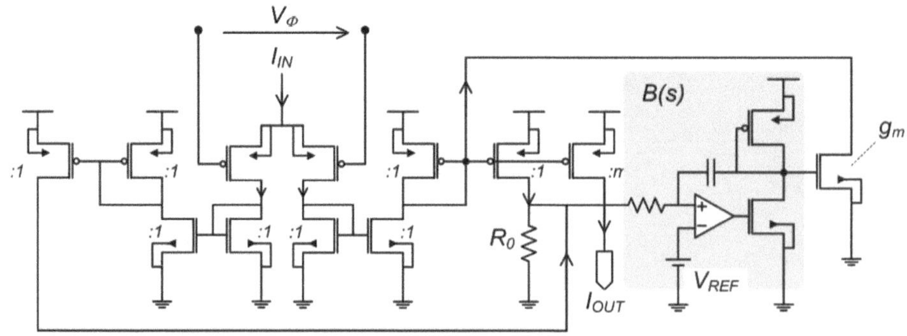

Fig. 3.6 Schematic of the reconfigurable base-band filter

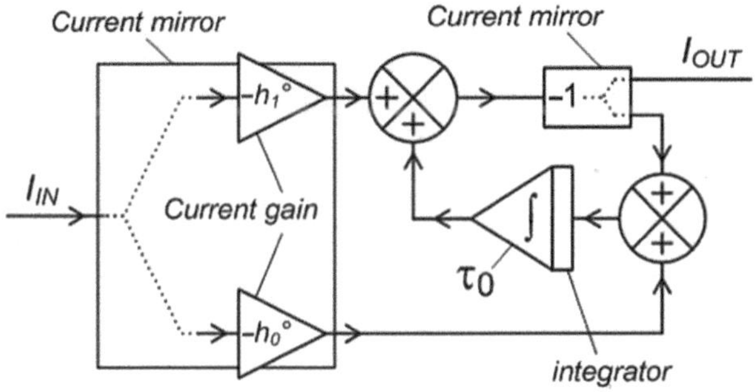

Fig. 3.7 Reconfigurable base-band filter synoptic

where

$$h_0^\circ(V_\phi) = \frac{1+f^\circ(V_\phi)}{2} \quad ; \quad h_1^\circ(V_\phi) = \frac{1-f^\circ(V_\phi)}{2} \quad ;$$

$$\omega_p = \frac{1}{\tau_0} \quad ; \quad \omega_Z = \omega_p \times \frac{h_0^\circ(V_\phi)}{h_1^\circ(V_\phi)} = \omega_p \times \frac{1+f^\circ(V_\phi)}{1-f^\circ(V_\phi)}$$

$$\forall V_\phi \in [V_{\phi\_min} \ ; \ V_{\phi\_max}], \ f^\circ(V_\phi) \in [-1 \ ; \ 1]$$

The function f° is determined by the input differential pair that is biased by the input current I_{IN} in Fig. 3.6. In convenience with the frequency response that is expected from the base-band filter, the zero ω_z can be shifted above/below the pole ω_p by reconfiguring the voltage V_Φ. The pole ω_p should be chosen consistently with the channel bandwidth and/or with the expected memory effects cut-off frequency (if estimated). Indeed, one of the goals of $H(s)$ is to combat the phase shift due to memory effects that generally results in the lower/upper IMD_3 asymmetry (see further Section 3.1.2.5 for details). In Fig. 3.8, the simulation of $H(s)$ transfer function is represented for various V_Φ values when ST Microelectronics 0.25 μm *BICMOS* technology is employed.

Several modes can be distinguished:

- For $V_\Phi < 0, f^\circ(0) > 0$, and $w_z > w_p$: $H(s)$ behaves as a low-pass filter and features a negative phase shift above the frequency ω_p. At the upper limit, $V_\phi \to V_{\phi\_min}$, we have $f^\circ(V_\phi) \to +1$ and $\omega_z \to +\infty$.
- When $V_\Phi = 0, f^\circ(0) = 0$, $w_z = w_p$, and $H(s)$ behaves approximately transparently.
- For $V_\Phi > 0, f^\circ(0) < 0$, and $w_z < w_p$: $H(s)$ behaves as a high-pass filter and features a positive phase shift above the frequency ω_p. At the lower limit, $V_\phi \to V_{\phi\_max}$, we have $f^\circ(V_\phi) \to -1$ and $\omega_z \to 0$.

The dual adaptive bias mechanism that is presented in Fig. 3.5 is modelled in the base-band frequency domain by the synoptic in Fig. 3.9.

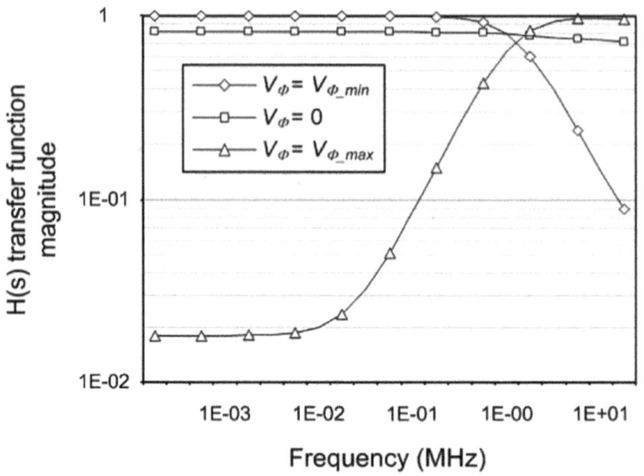

Fig. 3.8 Adjustable base-band filter response

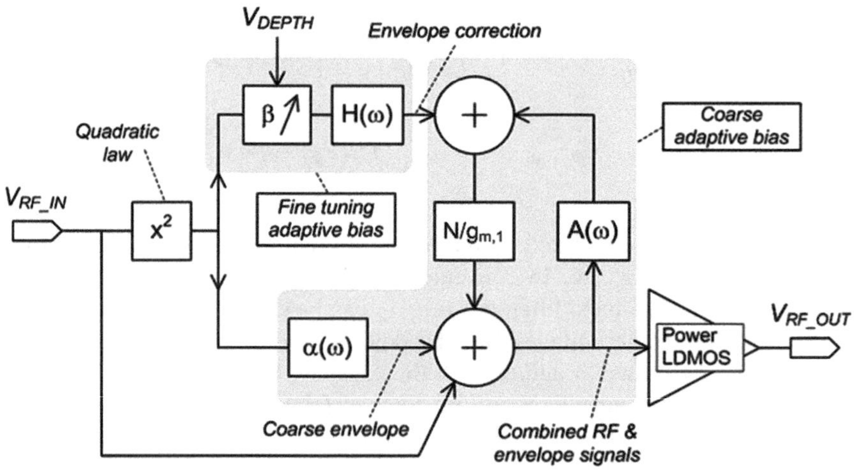

Fig. 3.9 Base-band system synoptic

The fundamental parameters of the adaptive system are:

- the *basic adaptability factor* $\alpha(\Delta\omega)$ that was introduced in (3.12)
- $\beta \times V_{DEPTH}$ represents the self-mixer conversion gain
- $H(\Delta\omega)$ represents the base-band filter transfer function,
- $A(\Delta\omega)$ represents the common-mode feedback transfer function,
- N represents the *LDMOS* mirror ratio,
- $g_{m,1}$ represents the transconductance gain of the power device M_{LDa}.

The closed-loop transfer function that links the consumed current with the *RF* envelope is inferred from Fig. 3.9:

$$\eta_{DEPTH}(\Delta\omega, V_{DEPTH}) = \frac{\partial^2 i_{dd}}{\partial v_g^2} = g_{m,1} \cdot \frac{\left(g_{m,1} \cdot \alpha(\Delta\omega) + N \cdot H(\Delta\omega) \cdot \beta \cdot V_{DEPTH}\right)}{g_{m,1} + N \cdot A(\Delta\omega)} \tag{3.16}$$

The factor η_{DEPTH} is the *adjustable adaptability factor*. From a physical point of view, η_{DEPTH} is the second derivate of base-band drain current I_{DD} with respect to an incremental variation in the *RF* gate voltage magnitude of the power *LDMOS*. Therefore it expresses to what extent the base-band drain current varies as a function of the input power. This term plays a central role in the *PA* adaptive behavior since it determines linearity, current consumption and efficiency.

Considering a dual-tone *RF* signal (ω_C, $\omega_C - \Delta\omega$), and according to Fig. 3.9, the base-band harmonics of the drain current can be expressed in frequency domain as a function of *coarse* and *fine tuning adaptive bias* parameters:

$$\begin{aligned} I_{DD}(\omega) = &\left(I_{DDQ} + \eta_{DEPTH}(0, V_{DEPTH}) \times \frac{\left|V_g\right|^2}{2}\right) \cdot \delta(\omega) \\ &+ \eta_{DEPTH}(\Delta\omega, V_{DEPTH}) \times \frac{\left|V_g\right|^2}{2} \cdot \delta(\omega - \Delta\omega) \end{aligned} \tag{3.17}$$

where I_{DDQ} is the (power-independent) quiescent bias current of a power *LDMOS*, and V_g is the *RF* input voltage magnitude.

3.1.2.4 Theory of Linearity Optimization via Dual Adaptive Bias

In the following section, the *non-linear current method* will be applied to the power transistor in order to formalize the effect of adaptive bias on linearity. Non-linear current sources are added to the classic small-signal schematic (Fig. 3.10). The term i_{g_nl} stands for the non-linear gate current sources that are mainly due to adaptive bias predistortion (see (3.5) and (3.8), and Fig. 3.4). The term i_{d_nl} stands for the non-linear drain current characteristic of the power device, and is expressed by means of $G_{m,j}$ transconductance terms.

An approximation of the transconductance gain non-linear components is inferred from (B.6) in the theoretical analysis of Appendix B:

$$G_{m,j}(\omega) \approx \frac{g_{m,j}}{\left(1 + j\,\omega\, C_{gd}\, R_L \frac{g_{m,1}}{Y_{g,1}(\omega)}\right)} \tag{3.18}$$

Based on (3.18) and considering a dual-tone *RF* signal (ω_1, $\omega_2 = \omega_1 - \Delta\omega$), the non-linear gate voltage harmonic at $\Delta\omega$ is expressed as follows:

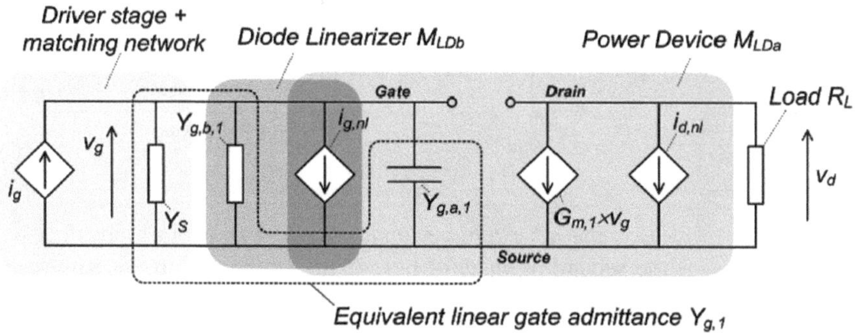

Fig. 3.10 Small signal schematic of the power transistor (non-linear sources included)

$$\begin{aligned}\frac{i_{g\_nl}(\Delta\omega)}{Y_{g,1}(\Delta\omega)} &= \frac{1}{2}\cdot\frac{\eta_{DEPTH}(\Delta\omega, V_{DEPTH})}{g_{m,1}}\cdot V_{g,\omega_1}\cdot V^*_{g,\omega_2}\\ &\approx \frac{1}{2}\cdot\frac{(g_{m1}\cdot\alpha(\Delta\omega)+N\cdot H(\Delta\omega)\cdot\beta\cdot V_{DEPTH})}{g_{m1}+N\cdot A(\Delta\omega)}\cdot V_{g,\omega_1}\cdot V^*_{g,\omega_2}\end{aligned}\tag{3.19}$$

Likewise, at the frequency $2\omega_1-\omega_2$ (>> the cut-off frequency of $A(\omega)$):

$$\frac{i_{g\_nl}(2\omega_1-\omega_2)}{Y_{g,1}(2\omega_1-\omega_2)}\approx\frac{3}{4}\cdot\kappa(\omega_1,\omega_1,-\omega_2)\cdot V^2_{g,\omega_1}\cdot V^*_{g,\omega_2}\tag{3.20}$$

The variations of the output drain voltage can be expressed in time domain as follows (taking into account linear and non-linear components):

$$\begin{aligned}v_d &= -R_L\cdot\left(\sum_{j=1}^{+\infty}G_{m,j}\cdot\left(v_g-\frac{i_{g\_nl}}{Y_{g,1}}\right)^j\right)\\ &= -R_L\cdot\left(G_{m,1}\cdot\left(v_g-\frac{i_{g\_nl}}{Y_{g,1}}\right)-\underbrace{\left(\sum_{j=1}^{+\infty}G_{m,j}\cdot\left(v_g-\frac{i_{g\_nl}}{Y_{g,1}}\right)^j\right)}_{i_{d\_nl}}\right)\end{aligned}\tag{3.21}$$

By substituting non linear components in (3.21) with their expressions (3.18), (3.19), and (3.20), and via proper development in frequency domain, the *RF* drain voltage harmonics can be obtained. The fundamental tone and the adjacent 3rd order inter-modulation product are given, respectively, by (3.22) and (3.23), at low power level:

$$V_d(\omega_1) = -R_L\cdot G_{m,1}(\omega_1)\cdot V_{g,\omega_1}+\circ\left(\left|V_{g,\omega_1}\right|\right)\tag{3.22}$$

and

$$V_d\left(2\omega_1-\omega_2\right)=-R_L\cdot G_{m,1}\left(\omega_1\right)\cdot V_{g,\omega_1}^2\cdot V_{g,\omega_2}^*\cdot\varepsilon\left(\omega_1,-\omega_2\right)+\circ\left(\left|V_{g,\omega_1}\right|^3\right) \tag{3.23}$$

where ε is a complex 3rd order function that expresses the *PA* non-linearity and that is given by:

$$\varepsilon\left(\omega_C,\Delta\omega-\omega_C\right)=\frac{1}{4}\times\left(\begin{array}{l}\underbrace{3\left(\frac{g_{m,3}}{g_{m,1}}+\kappa\left(\omega_C,\omega_C,\Delta\omega-\omega_C\right)\right)}_{3rd\ degree\ non-linearity}\\+\underbrace{\frac{g_{m,2}}{g_{m,1}^2}\cdot\eta_{DEPTH}\left(\Delta\omega,V_{DEPTH}\right)}_{compound\ 3rd\ order\ non-linearity}\end{array}\right) \tag{3.24}$$

Two kinds of non-linear contributions can be distinguished in the expression $\varepsilon(\omega_C, \Delta\omega\text{–}\omega_C)$:

- 3rd degree non-linearities that *directly* generate 3rd order inter-modulation products. They result in complex and frequency-independent inter-modulation products over a *TX* channel due to the relatively narrow channel bandwidth compared with the carrier frequency ($BW_{channel} = 5$ MHz << Carrier frequency = 1.95 GHz). Consequently, the upper and lower contributions of κ are roughly identical:

$$\kappa\left(\omega_C,\omega_C,\Delta\omega-\omega_C\right)\approx\kappa\left(\omega_C-\Delta\omega,\omega_C-\Delta\omega,-\omega_C\right)$$

- *Compound* 3rd order non-linearities that most contribute to memory effects. They are caused by the cascaded combination of several non-linearities and are highly frequency-dependent due to the intermediate step in the base-band frequency domain.

The output intercept point is determined by the function ε as outlined by the following interception condition:

$$\left|\frac{V_{d,\,2\omega_1-\omega_2}}{V_{d,\,\omega_1}}\right|_{@IP_3}=1\ \Leftrightarrow\ \left|V_{g,\omega_1}\right|_{@IP_3}^2=\left|\varepsilon\left(\omega_1,-\omega_2\right)\right|^{-1} \tag{3.25}$$

Substituting (3.18) in (3.25) results in:

$$OIP_3 = \frac{\left|V_{d,\omega_1}\right|^2_{OIP_3}}{2\,R_L} = \frac{R_L \times \left|G_{m,1}(\omega_1)\right|^2 \times \left|V_{g,\omega_1}\right|^2_{@IIP_3}}{2} = \frac{g_{m,1}^2 \cdot R_L \cdot |\,\varepsilon(\omega_1,-\omega_2)\,|^{-1}}{2\left|\left(1 + j\,\omega\,C_{gd}\,R_L\,\frac{g_{m,1}}{Y_{g,1}(\omega_1)}\right)\right|^2} \tag{3.26}$$

It is clear that a *PA* linearization technique consists in cancelling the term $\varepsilon(\omega_1,-\omega_2)$ for any power condition, and for any frequency spacing $|\Delta\omega|=|\omega_1-\omega_2|\in[0;\,BW_{channel}]$. The principle of this linearization is illustrated in Fig. 3.11. For sake of clarity, a unit input *RF* voltage is considered in this figure. The mixing products (a) represent the 3rd degree "direct" non-linearities. The mixing products (b) represent the base-band gate voltage harmonics that arise from adaptive bias. These harmonics are themselves convoluted with the input *RF* voltage to generate the "compound" 3rd order mixing products (c) (as opposed to the 3rd degree "direct" non-linearities (a)). According to (3.24), the expression $\varepsilon(\omega_1,-\omega_2)$ can be cancelled if $\eta_{DEPTH}\,(\Delta\omega, V_{DEPTH})$ equals an optimum 3rd order complex function $\eta_{OPT}\,(\omega_1,\omega_2,-\omega_2)$ that is written as follows:

$$\eta_{OPT}\,(\omega_1\,,\,\omega_1,-\omega_2) = -3\cdot\frac{g_{m,1}^2}{g_{m,2}}\cdot\left(\kappa\,(\omega_1\,,\,\omega_1,-\omega_2) + \frac{g_{m,3}}{g_{m,1}}\right) \tag{3.27}$$

In practice, the cancellation condition must be fulfilled over the whole complex plan. Stating that κ is essentially an imaginary function, the cancellation condition on the real part is inferred from (3.27).

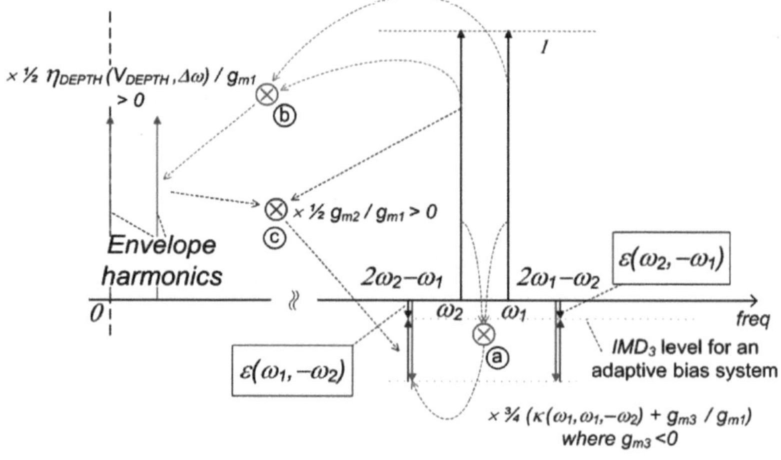

Fig. 3.11 Illustration of adaptive bias linearization in frequency domain

$$\Re\left(\eta_{DEPTH}\left(\Delta\omega, V_{DEPTH}\right)\right) = \Re\left(\eta_{OPT}\left(\omega_1, \omega_1, -\omega_2\right)\right) \approx -3\, g_{m,1} \times \frac{g_{m,3}}{g_{m,2}} \tag{3.28}$$

For a narrow channel bandwidth application, the condition of linearity optimization can be expressed in terms of V_{DEPTH} by substituting (3.16) in (3.28):

$$V_{DEPTH} \approx -\frac{\left(g_{m,1} \cdot \alpha(0) + 3 \times \left(g_{m,1} + N \cdot A(0)\right) \times \frac{g_{m,3}}{g_{m,2}}\right)}{N \cdot H(0) \cdot \beta} \tag{3.29}$$

Figure 3.12 gives the waves $\Re(\eta_{OPT})$ and $\Re(\eta_{DEPTH})$ for various V_{DEPTH} values. At a specific power level (i.e. bias level), the condition $\Re(\eta_{DEPTH}) = \Re(\eta_{OPT})$ is fulfilled for a specific V_{DEPTH} value. This is the intercept point of both waves. Below this optimum point, the non linear term due to envelope injection exceeds the intrinsic non linearity of the power stage, resulting in envelope *over-injection*. Above the optimum point, *under-injection* is observed.

In order to prevent over(under)-injection and keep linearity as high as possible at any average power level, it is desirable to shift adaptively and consistently the η_{depth} waveform upwards or downwards by means of V_{DEPTH}. For an optimum response, V_{DEPTH} voltage is to be reconfigured as a function of the average power level and of the addressed standard, e.g. via a look-up-table containing a control signature. Therefore, contrary to most adaptive architectures, this bias system can be considered as a 2nd order system in respect to power.

Figure 3.13 is a vector illustration of the complex non linear term ε. This vector representation distinguishes between the various non-linear contributions that can be

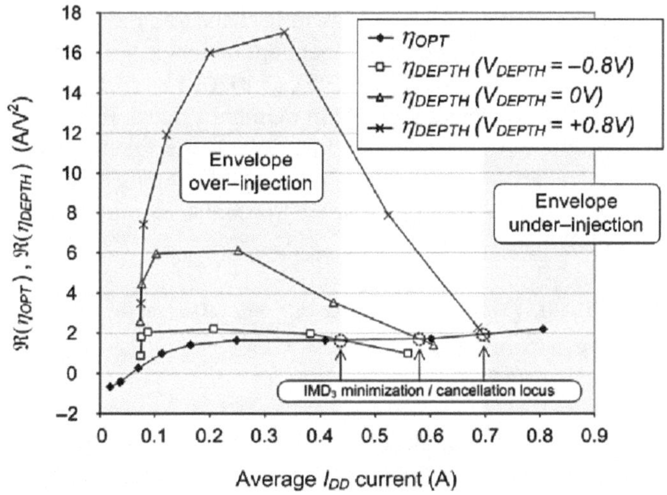

Fig. 3.12 Illustration of linearity optimization by the proper envelope injection via V_{DEPTH} (interception points represent the optimal locus as a function of bias/power level)

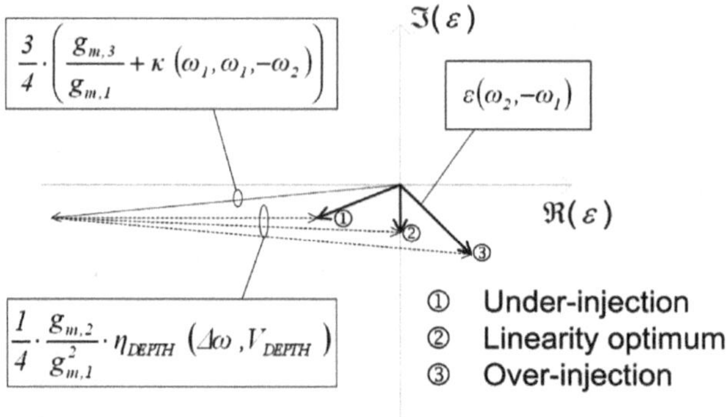

Fig. 3.13 Vector illustration of linearity improvement via V_{DEPTH} optimization

found in (3.24). The vector ① (resp. ③) depicts a case of envelope under-injection (resp. over-injection). The vector ② depicts a case of quasi-optimum envelope injection (i.e. $|\varepsilon|$ is minimized).

3.1.2.5 Memory Effects and Adaptive Bias

Extensive analyses have been carried out on memory effects in power amplifiers (Vuolevi 2001, Boumaiza and Ghannouchi 2003, Brinkhoff 2004, Kimura et al. 2004, Liu et al. 2005, Aikio and Rahkonen 2005, Leyssenne et al. 2006). Strong memory effects are generally diagnosed when significant asymmetry is observed between lower and upper inter-modulation products. They can have several physical roots. Some electrical parameters (such as the 2nd order capacitances of the power devices $C_{gs,2}$ or $C_{gd,2}$) have an influence on the frequency dependence of the 3rd order gain over the channel bandwidth. This was more particularly highlighted by the analysis of the frequency-dependent factors $\eta_{DEPTH}(\Delta\omega, V_{DEPTH})$ and $\varepsilon(\omega_C, \Delta\omega\text{–}\omega_C)$ in previous sections (see (3.5), (3.8), (3.11), (3.16), and (3.24)). The temperature variation also plays a key role in memory effects. It is a 2nd order function with regard to the *RF* magnitude and modulates the *PA* linear parameters via the thermal dependencies ($\partial\, g_{m,1}/\partial\, T°$, and $\partial\, y_{g,1}/\partial\, T°$ amongst others). In the case of a dual-tone signal and based on (3.17), the thermal spectrum of an adaptive-bias *PA* is given by (3.30):

$$\Delta T(\omega) = \frac{|V_g|^2}{2}\cdot\begin{pmatrix} Z_{TH}(0)\cdot\left(V_{DD}\cdot\eta_{DEPTH}(0, V_{DEPTH}) - R_L\cdot\left|G_{m,1}(\omega_C)\right|^2\right)\cdot\delta(\omega) \\ +Z_{TH}(\Delta\omega)\cdot\left(V_{DD}\cdot\eta_{DEPTH}(\Delta\omega, V_{DEPTH}) - R_L\cdot\left|G_{m,1}(\omega_C)\right|^2\right)\cdot\delta(\omega-\Delta\omega)\end{pmatrix} \tag{3.30}$$

where Z_{TH} is the *PA* thermal impedance, V_{DD} is the power supply, V_g is the *RF* magnitude, $G_{\mathrm{m},1}$ is the *PA* complex linear transconductance gain, R_{L} is the output load and η_{DEPTH} is the adaptability factor.

An adaptive bias *PA* is therefore a potentially sensitive circuit to memory effects. A figure of merit of a reconfigurable *PA* is its capability to control upper/lower

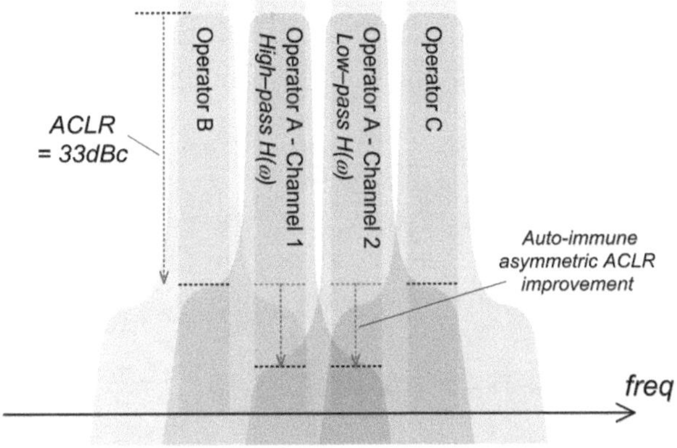

Fig. 3.14 Adaptive IMD_3 unbalance and mutual immunization of adjacent channels

asymmetry. This can be carried out by adaptively adjusting the frequency-dependent phase shift due to the base-band filter *H(f)* (see the synoptic in Fig. 3.5). In some specific cases, upper/lower asymmetry can be desirable. For example, mobile phone operators sometimes own adjacent channels and they may wish to privilege selectively asymmetric *ACLR* in order to improve mutual immunity of their own channels (Fig. 3.14). The base-station that receives the signals from the users of channel1 and channel2 (phone operator A) will be able to discriminate them more easily since the noise level is reduced by the amount of the lower/upper *ACLR* unbalance, whereby the *Quality-of-Service* and the affordable number of users are enhanced.

Such asymmetry can be understood via a vector illustration of the non-linear function ε in Fig. 3.15. It is assumed that two *RF* tones ω_1 and ω_2 are applied to a *PA* with a spacing $\Delta\omega = \omega_1 - \omega_2$. The various non-linear components that contribute to the generation of intermodulation products are depicted by separate vectors. Dynamic temperature modulation can be of great importance and is modeled by the vector $M(\pm\Delta\omega)$. The frequency-dependent adaptive contribution $\eta_{DEPTH}(\Delta\omega, V_{DEPTH})$ is a memory effect on its own and may be configured to compensate for $M(\pm\Delta\omega)$. The 3rd degree non linear components (that are related to the factors g_{m3}, and κ) are close to being frequency independent over the *RF* bandwidth (i.e. $\kappa(\omega_2,\omega_2,-\omega_1)\sim\kappa(\omega_1,\omega_1,-\omega_2)$). Their contributions to upper/lower IMD_3 are therefore represented by merged vectors. In contrast, the 2nd-order non-linear components that are related to the frequency spacings $+\Delta\omega$ and $-\Delta\omega$ are conjugate one of the other, i.e. $\eta_{DEPTH}(+\Delta\omega) = \eta_{DEPTH}{}^{*}(-\Delta\omega)$ and $M(+\Delta\omega) = M^{*}(-\Delta\omega)$. This explains that the resulting vectors ε $(\omega_2,-\omega_1)$ and ε $(\omega_1,-\omega_2)$ may have different magnitude and phase. Severe IMD_3 unbalance can be encountered when the phase shift of $\eta_{DEPTH}(\pm\Delta\omega)$ and $M(\pm\Delta\omega)$ do not compensate for each other. This is especially true for broad frequency spacings. The phase shift polarity of the memory effects determines the direction of IMD_3 unbalance.

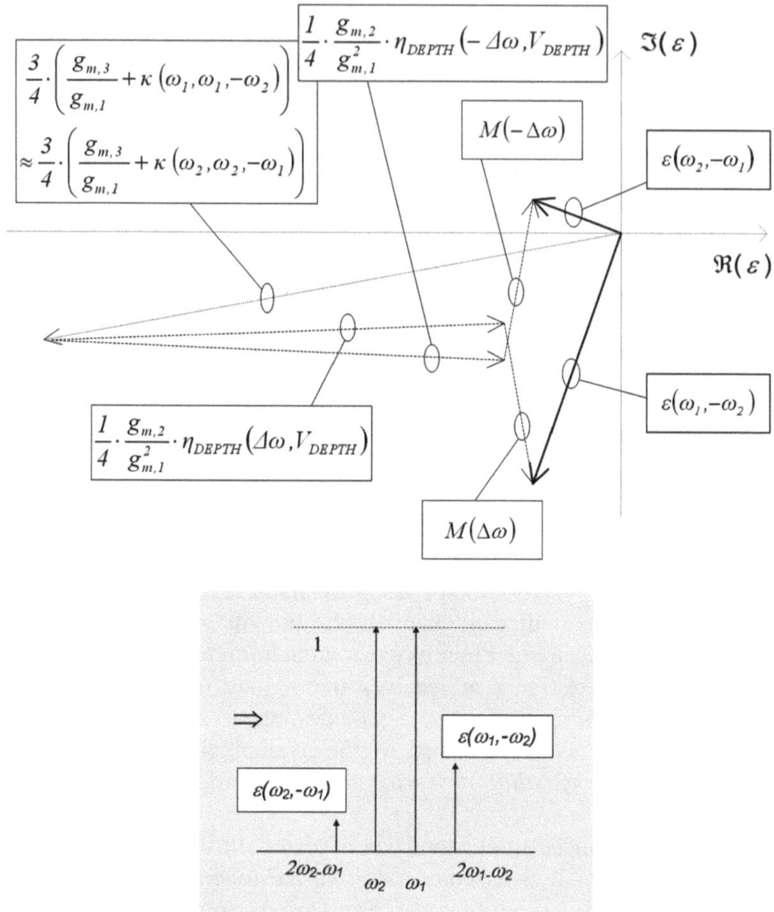

Fig. 3.15 Vector explanation of IMD_3 unbalance

3.2 Design and Measurement of the Integrated Passive Device Dedicated to *PA* Module

3.2.1 MOBILIS IPD Design

The *IPD* passive network transforms the output single-ended 50 Ω load into a 100 Ω differential impedance ($2 \times R_{BALUN}$), then into a 5 Ω differential impedance (*IPD* input). The synoptic of this circuit is depicted in Fig. 3.16. The *IPD* was designed stand-alone under the constraint of low loss and wide bandwidth covering *DCS* and *WCDMA* operating bands [1.71, 1.98 GHz].

So as to reduce the die area used, the *IPD* was limited to a 4th order network. The higher the network order, the shorter the distance that the impedance locus goes through on the Smith chart (i.e. the broader the bandwidth). The *IPD* bandwidth

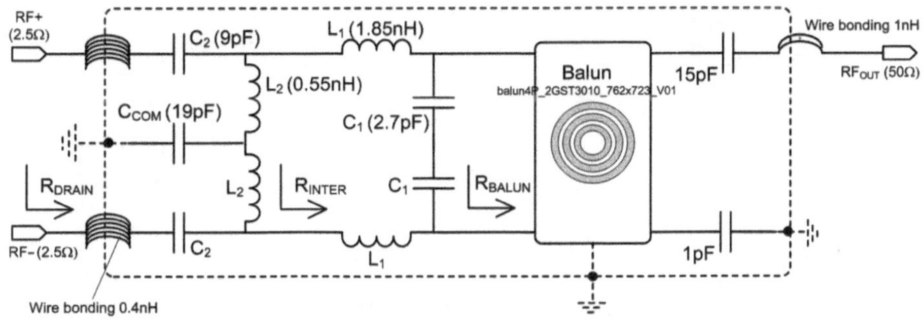

Fig. 3.16 *IPD* schematic

is maximized by combining a high-pass (DC blocker on silicon side) with a low-pass network (on balun side) and by fulfilling the conditions (3.31). The resistance R_{INTER} is the mid-point impedance level. To improve linearity, the out-of-band *PA* termination was addressed. To this end, the high-value C_{COM} capacitance minimizes the 2nd order (common-mode) input *IPD* impedance at twice the carrier frequency. In this way, output voltage peaks that might drive the *PA* into excessive saturation are avoided, as there could be a possible destruction of the *PA*. An optimization was then carried out taking into account the *IPD* device models. Two *IPDs* were taped-out: one for the purpose of assembly (Fig. 3.17), and another for on-wafer measurements (Fig. 3.18).

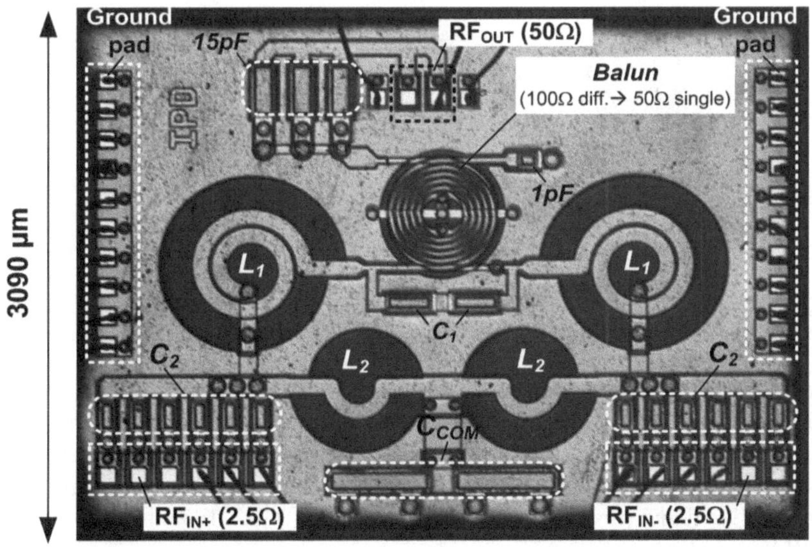

Fig. 3.17 Assembly-dedicated *IPD* layout

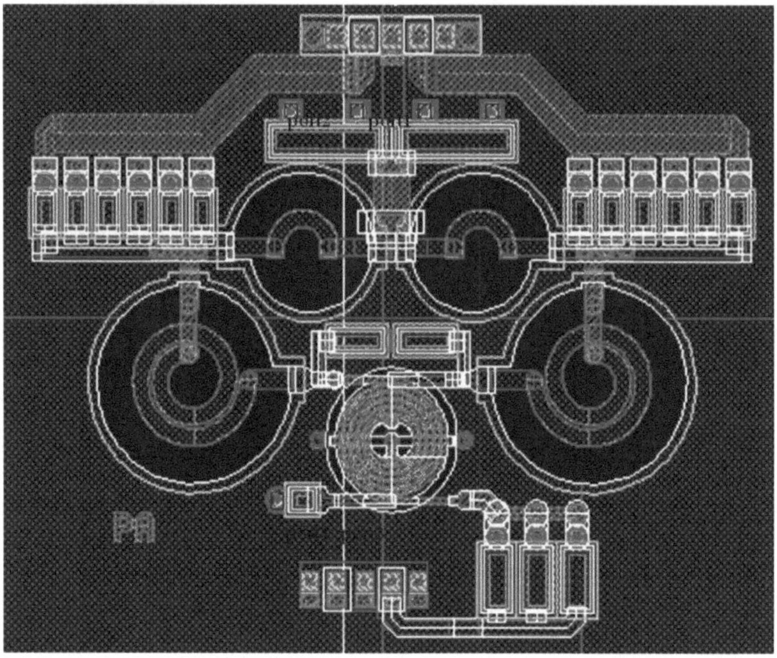

Fig. 3.18 On-wafer-dedicated *IPD* circuit

$$\begin{cases} R_{inter} = (R_{balun} \cdot R_{drain})^{1/2} \\ L_1 = \dfrac{1}{4\pi^2 f_{RF}^2 \cdot C_1} = \dfrac{R_{balun}^{3/4} \times R_{drain}^{1/4}}{2\pi f_{RF}} \approx 1.88\text{nH} \\ L_2 = \dfrac{1}{4\pi^2 f_{RF}^2 \cdot C_2} = \dfrac{R_{balun}^{1/4} \times R_{drain}^{3/4}}{2\pi f_{RF}} \approx 0.45\text{nH} \end{cases} \quad (3.31)$$

where $R_{balun} = 50\ \Omega$, $R_{drain} = 3\ \Omega$, $f_{RF} = 1/2 \times (1.71 + 1.98)$ GHz

3.2.2 IPD Characterization

The *IPD* circuit was characterized by means of on-wafer measurement according to a dedicated *IPD* topology circuit (Fig. 3.18). The measured 1.85 dB insertion loss is slightly higher than the 1.1 dB simulated insertion loss, due to the non de-embedding of the microstrip differential access (port1). The out-of-band rejection at 0.9 GHz in the *GSM900* band (resp. at 2.6 GHz) is 27 dB (resp. 8.3 dB), in compliance with *MOBILIS* specifications.

Figure 3.19 presents minor input (5 Ω differential port) and output (50 Ω differential port) return losses, below –12 dB in the [1.7 GHz ; 2 GHz] band, with two

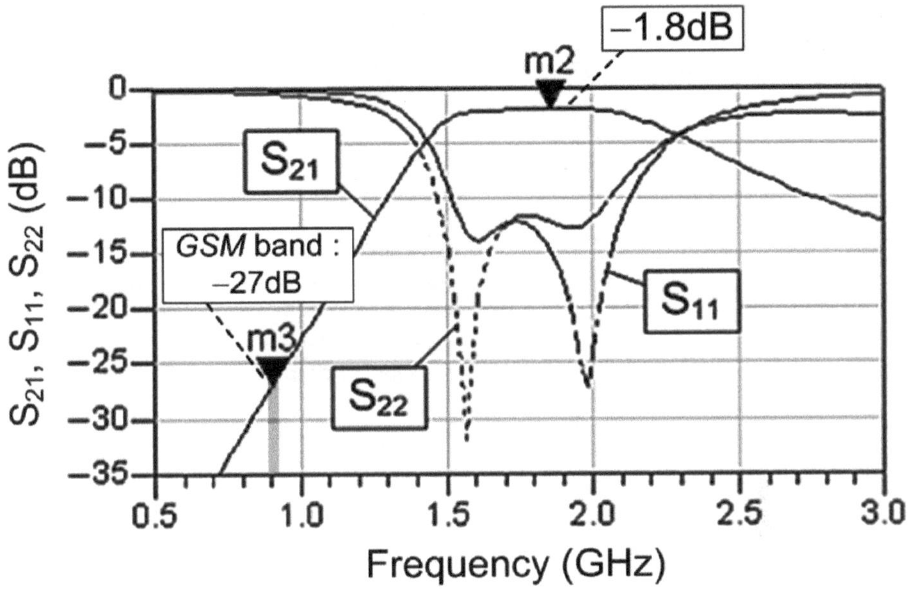

Fig. 3.19 Measured *IPD* *S*-parameters

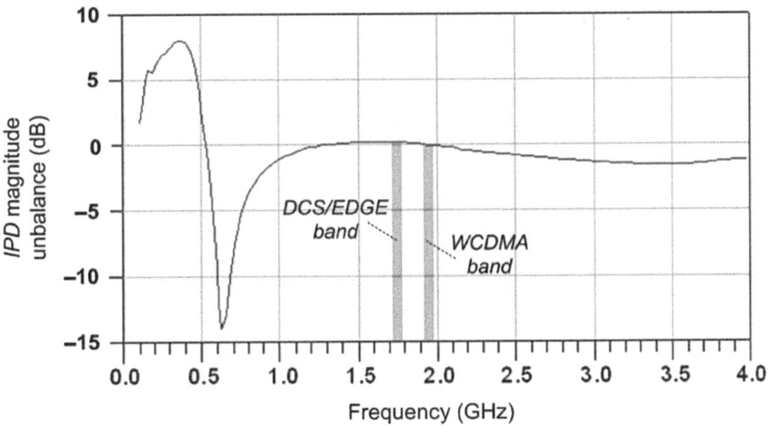

Fig. 3.20 Measured *IPD* magnitude imbalance

zeroes at 1.6 and 2 GHz. A 0.1 dB magnitude imbalance (Fig. 3.20) and 1° phase imbalance were observed on the differential port.

Three-port measurements are given in Fig. 3.21. Ports 1 and 2 are 2.5 Ω single-ended ports (*PA* silicon side), whereas port 3 is the duplexer antenna 50 Ω port. The comparison with the simulation presents somewhat reduced bandwidth and –4.85 dB transfer gains (S_{32} and S_{31}) instead of the simulated –3.95 dB.

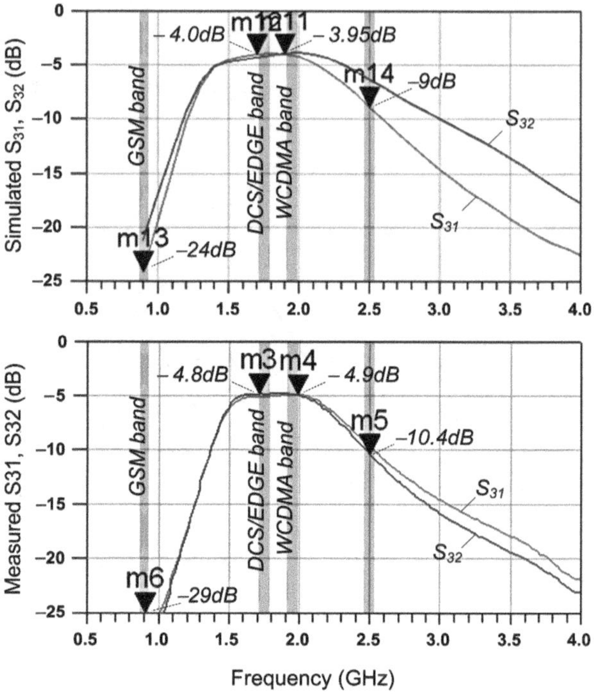

Fig. 3.21 Measured and simulated transfer S-parameters S_{32}, and S_{31}

3.3 Design and Simulation of the Adaptive Bias Silicon *PA*

3.3.1 PA Silicon Design

Most commercial Li-Ion batteries feature an output voltage in the range of 3.4–3.6 V. Assuming the power supply is 3.5 V, and the targeted output power is 30 dBm (i.e. 1 W), the output differential load (drain to drain) can be estimated as follows:

$$R_{L\_diff} = \frac{(V_{DD} - V_{headroom})}{P_{OUT}} \approx (3.5 - 1)^2 \approx 6\,\Omega_{diff} \tag{3.32}$$

The 1 V headroom voltage represents the margin on drain voltage that must be taken into account so as to keep the power *LDMOS* out of the triode region.

The *PA* module that is presented here is designed according to the architectures in Figs. 3.1 and 3.5. The power devices (M_{LDa} in Fig. 3.5) are designed as 336×20 μm-wide 2nd generation *LDMOS*. Such device size results in a 1.5 siemens transconductance gain at the bias level for which transition frequency is maximum (V_{GS0} = 1.25 V, see Figs. 1.29 and 1.30). The ratio $C_{gs,1}/C_{gd,1}$ is approximately 5 (see Fig. 1.31). The real part of the output admittance is therefore:

$$\Re\left(Y_{out}\right) = g_{m,1} \times \frac{C_{gd,1}}{C_{gs,1}} \approx 0.3S \rightarrow \Re\left(Y_{out}\right)^{-1} \approx 3\,\Omega \tag{3.33}$$

This enables good impedance matching with the 3 Ω single-ended load (i.e. 6 Ω differential load) that is required by the condition on power capability (3.32). At lower bias (i.e. at lower power level), the real part of Y_{OUT} is reduced and the matching condition is not so fully fulfilled which is a drawback of adaptive bias *PA* architectures.

The *linearizer* devices (M_{LDb} in Fig. 3.5) and their ballast resistors r_g are designed wide enough to provide a sufficient analog predistortion to power devices (in magnitude and phase). They are sized as 48 × 20 μm-wide 2nd generation *LDMOS*. The soft feedback network $A(s)$ that intervenes in the adaptive bias system is designed as a low-pass network with a 10 MHz pole and features a transconductance gain of 13 mS. The driver stage is built on the same topology as the power stage with a 1:5 ratio. The driver output impedance of approximately 40 Ω is matched to the ~10 Ω power stage input impedance by means of a high-pass LC network (0.8 nH shunt inductor, 12 pF series capacitor).

Stability is a critical issue in the design of power amplifiers that is generally dealt through the analysis of stability circles. These circles represent the various load values for which instability may arise. Unconditional stability (i.e. that is load independent) is a desirable condition but can hardly be fulfilled in practice when other conditions over gain, linearity and efficiency are specified. Appendix B and more particularly Appendix D jointly expose the practical method that was employed to minimize the risks of instability. The trade-off stability/efficiency/gain is coped with by the proper design of the *diode linearizer* that plays the role of a shunt resistance and by loosening the impedance matching condition between the driver and power stages.

3.3.2 Simulations

3.3.2.1 Small-Signal Performance of the *PA* Module

The *S*-parameters of the *PA* module (silicon+*IPD*) were simulated with Agilent *Advanced Design System*. The overall transfer function S_{21} is depicted in Fig. 3.22. The silicon and *IPD* are connected by 6 bonding wires. This interface is modelled by a 0.5 nH parasitic series inductor that was taken into account in the *IPD* design. Two cases are considered in this paragraph. Firstly, component models only are used for the simulation (white squares in Fig. 3.22). The *PA* module gain is 27.5 dB in the *DCS* band and 26 dB in the *WCDMA* band. The *PA* module 3 dB bandwidth is found to be 500 MHz.

Secondly, a co-simulation is carried out (dark squares) by modelling the *IPD* as a four-port touchstone file (.s4p) and employing the data that was previously extracted from *IPD* characterization. In this case, the result is consistent with the exclusively model-based simulation but bandwidth is reduced and power gain is attenuated by 0.6 dB.

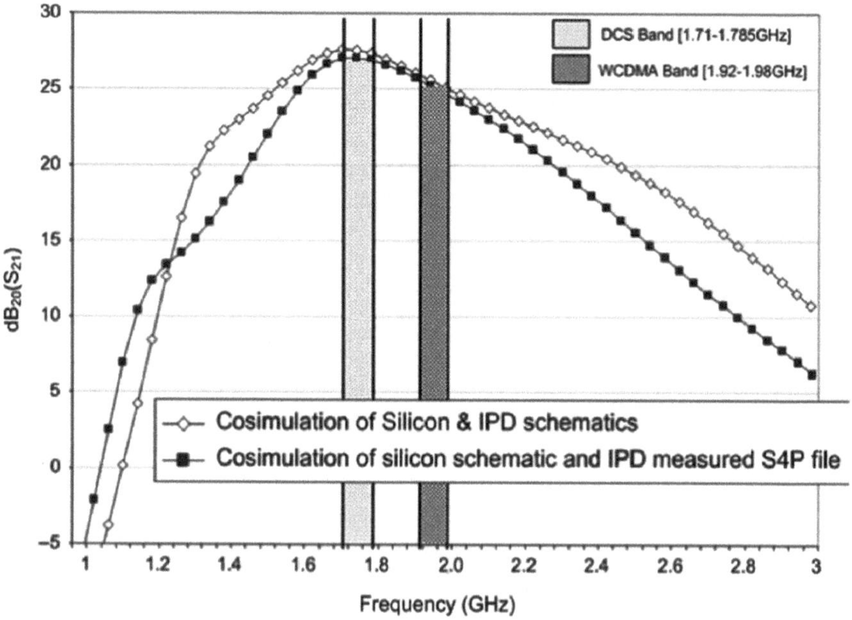

Fig. 3.22 Simulated transfer *S*-parameters S_{21} with *IPD* schematic and *IPD* actual response (measured .s4p)

3.3.2.2 Noise Performance of the *PA* Module

As explained in Chapter 1, *PA* noise specifications are not to be neglected especially in the *RX* band for a full duplex standard such as *WCDMA*. The *PA* module output noise density was simulated in Cadence. Using a small signal noise analysis (*noise* + *S-parameters*), it was found to be 7 nV/sqrt(Hz) at 2.14 GHz, i.e. –153 dBm/Hz (Fig. 3.23).

Noise response is bias dependent, and consequently it depends on the *RF* power level for an adaptive *PA*. That is the reason why a large-signal noise analysis was carried out (*pnoise* + *pseudo-steady-state*) at a high power level. In this case, *PA* module output noise density is 9.3 nV/sqrt(Hz) at 2.14 GHz, i.e. –150 dBm/Hz, which remains compliant with commonly used specifications in the *PA* market.

3.3.2.3 Continuous-Wave Performance of the *PA* Module

The simulated *PA* module compression point was found to be 31 dBm (Fig. 3.24). By modulating V_{DEPTH} voltage, a slight inflexion in P_{OUT} vs. P_{IN} is observed.

As demonstrated in Section 3.1.2.2, adaptive bias introduces an inflexion in the *AM/PM* response. This inflexion can be modulated by means of the voltage V_{DEPTH} (Fig. 3.25). More particularly, the power level for which the reversal of *AM/PM* polarity is observed can be shifted on purpose. By adaptively setting V_{DEPTH} as a function of the power level (from a low V_{DEPTH} value at low power level to a high

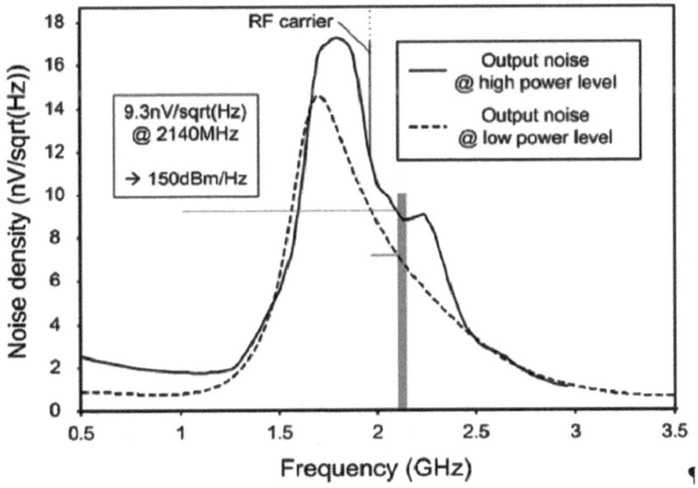

Fig. 3.23 Noise power density at *PA* silicon output

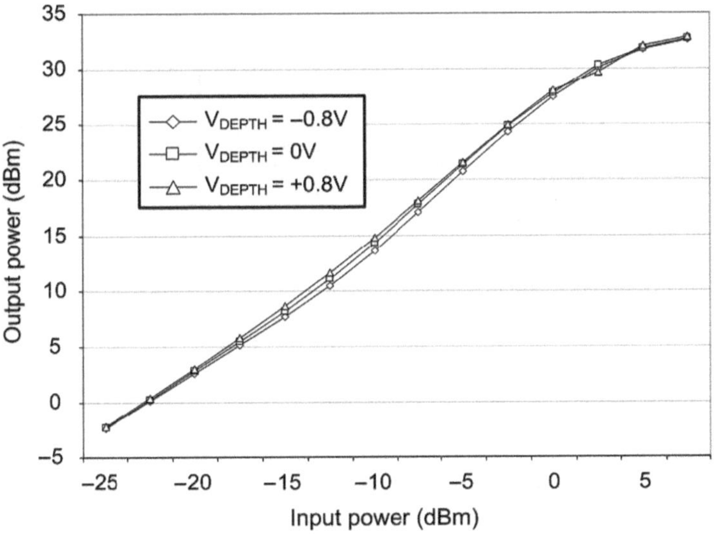

Fig. 3.24 P_{OUT} vs. P_{IN} for various V_{DEPTH} values

V_{DEPTH} value at high power level), it is possible to interpolate the various characteristics in Fig. 3.25 and to flatten the *finely tuned AM/PM* response over a broad power range. Power added efficiency was simulated in different load conditions (Fig. 3.26). As expected, the maximum *PAE* is up to 42% when silicon is stand-alone (ideal 6 Ω differential load). The co-simulation of silicon and *IPD* schematics presents a

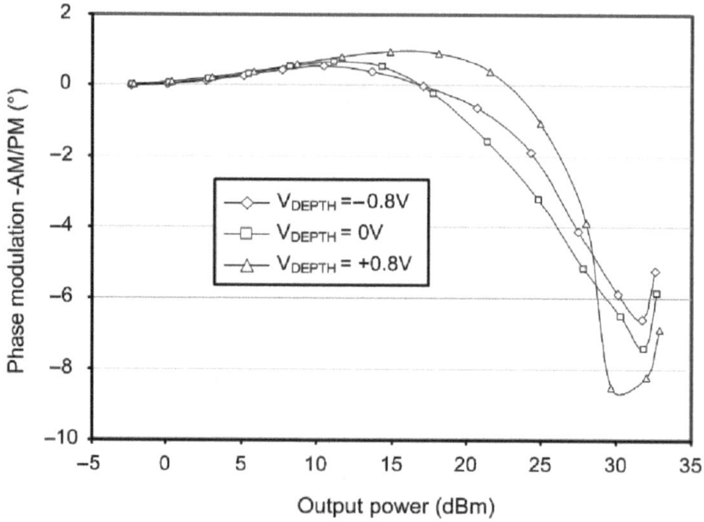

Fig. 3.25 Output phase modulation vs. P_{OUT} (*AM/PM*) for various V_{DEPTH} values

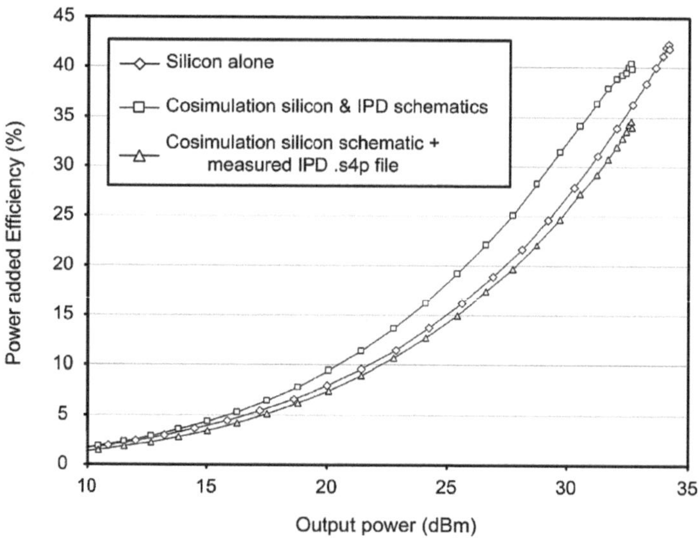

Fig. 3.26 Power added efficiency vs. P_{OUT} in various load condition

maximum power added efficiency of 40%. In contrast, the co-simulation of silicon schematic and *IPD* .s4p file presents reduced *PAE* performance (down to 35%). Thanks to adaptive bias, *PAE* at the medium power level $P_{OUT} = 21$ dBm (=10 dB back-off) remains above 8%.

3.3.2.4 Dual-Tone Performance of the *PA* Module

The *PA* module dual-tone performance was analyzed by means of harmonic balance simulations (*ADS* simulator) for various V_{DEPTH} values and assuming a 5 MHz-spacing (Fig. 3.27). In this example, *H* filter is replaced with a 30 MHz single-pole low-pass filter.

As expected, at a given power level, IMD_3 can be maximized by choosing the most appropriate V_{DEPTH} values. For example, at $P_{IN} = -25$ dBm, IMD_3 is up to 58 dBc with $V_{DEPTH} = -0.6$ V. At $P_{IN} = -18$ dBm, IMD_3 is up to 47 dBc with $V_{DEPTH} = -0.9$ V. At $P_{IN} = -7$ dBm, IMD_3 is up to 25 dBc with $V_{DEPTH} =$ 0.9 V. The interpolated locus of maximal IMD_3 is represented by the dotted line in Fig. 3.27 and can be obtained by adaptively configuring the V_{DEPTH} voltage as a function of the average power level.

In a similar manner in the complex plan, the normalized non-linear 3rd order kernel G_3/G_1 of the *PA* module was also extracted via harmonic balance simulations using a 5 MHz-spaced dual-tone signal, for various power levels and V_{DEPTH} values (Fig. 3.28). Arrows indicate the trend towards increasing power levels. It is worth noting that an increase in V_{DEPTH} results in a clockwise rotation of the G_3/G_1 locus. As stated about IMD_3, the interpolated locus of the minimal G_3/G_1 modulus (bold dark line in Fig. 3.28) can be obtained by adaptively configuring the V_{DEPTH} voltage as a function of the average power level.

The bandwidth of the adaptive bias mechanism (i.e. the cut-off frequency of the factor $\eta_{DEPTH}(\Delta\omega, V_{DEPTH})$ as defined in (3.18)) is an important feature since it determines if the envelope tracking is compliant with modern fast-rate standards. In

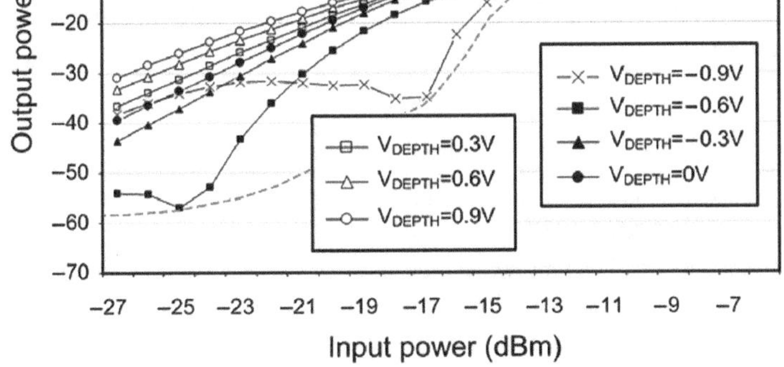

Fig. 3.27 IMD_3 for various V_{DEPTH} values

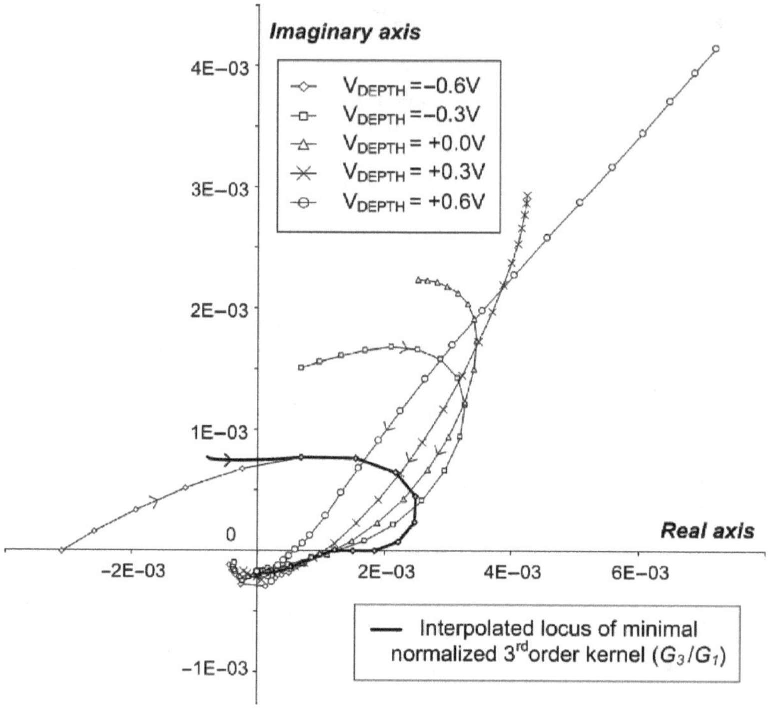

Fig. 3.28 Normalized 3rd order kernel for various V_{DEPTH} values

order to characterize this adaptive bias, harmonic balance simulations were carried-out for various V_{DEPTH} values. A dual-tone *RF* signal with a fixed power level is applied to the *PA* module and its spacing is swept. The base-band gate voltage magnitude of the power device was extracted and normalized by its value at $V_{DEPTH} = 0$ V and quasi-zero spacing (Fig. 3.29).

The relative gate magnitude deviation due to the adaptive bias is as high as ±40% (from a ratio of 0.5 to a ratio of 1.4). For a purpose of compliance with *WCDMA*, for which channel bandwidth is 5 MHz, the soft feedback network $A(s)$ was designed in such a way that the envelope tracking bandwidth rises above 10 MHz.

3.3.2.5 *PA* Module Performance with a *HPSK* Modulated Signal

The linearity performance of the *PA* module was characterized by Envelope simulations with a 3.84 Mcps *HPSK* modulated signal (uplink *WCDMA*) for various V_{DEPTH} values. Figure 3.30 shows the spectra of the *RF* output signal at 1.95 GHz (see top left image) and of the base-band consumed current I_{DD} (see bottom left image). The time domain section of Fig. 3.30 shows that the variations of the consumed current I_{DD} (see bottom right image) dynamically track those of the *RF* envelope (see top right image).

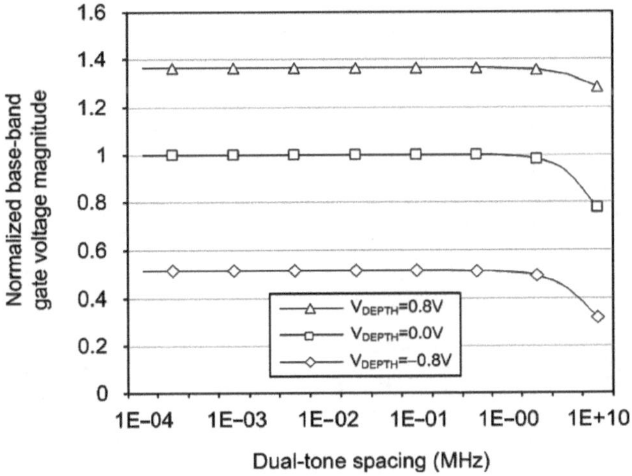

Fig. 3.29 Adaptive bias frequency bandwidth for various V_{DEPTH} values

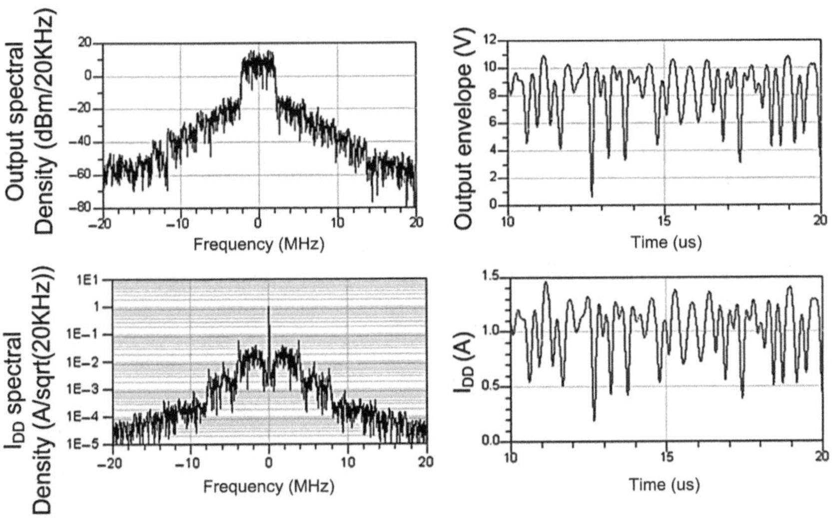

Fig. 3.30 Adaptive bias illustration with a 3.84 Mcps *HPSK* modulated signal @$P_{OUT} = 28$ dBm, $ACLR = 33$ dBc, both in frequency and time domains

The *Adjacent Channel Leakage Ratio* and the *RMS* value of *Error Vector Magnitude* were extracted by post-processing over a time slot with the software *ADS*. The fundamental benefit of the proposed topology is that, at a given average power level, *ACLR* (resp. *EVM*) can be maximized (resp. minimized) by choosing the most appropriate V_{DEPTH} value (Figs. 3.31 and 3.32).

Indeed, adaptive-bias power amplifiers generally present typical *ACLR*/*EVM* modulations over the power range. The proposed architecture makes it possible to shift the optimum *ACLR* peak and/or *EVM* notch gradually.

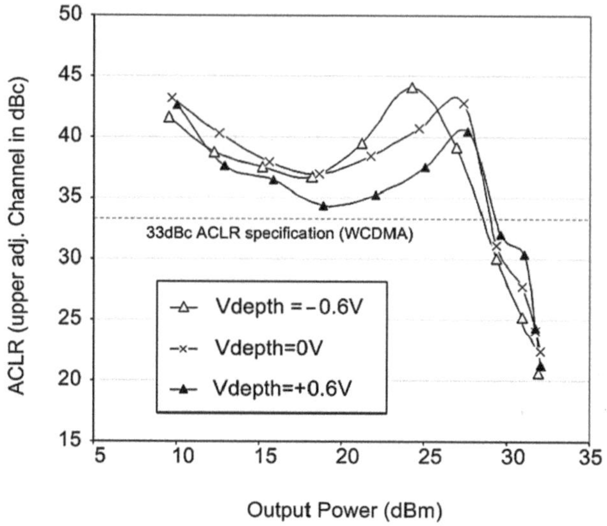

Fig. 3.31 *ACLR* vs. average output power (*WCDMA*) for various V_{DEPTH} values

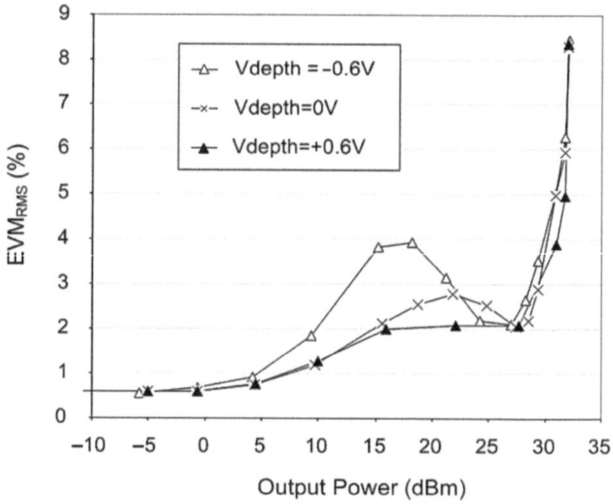

Fig. 3.32 *EVM* vs. average output power (*WCDMA*) for various V_{DEPTH} values

The maximum output power in compliance with the 33 dBc *ACLR WCDMA* specification is $P_{OUT} = 29$ dBm, which only implies a 2 dB back-off with regard to OCP_1 (see Fig. 3.24). The maximum output power in compliance with a 5% EVM_{RMS} (industrial *state of the art*) is $P_{OUT} = 31$ dBm (no back-off). This reduction of the necessary back-off (in comparison with the 3 dB *PAPR* that is commonly considered) is an important benefit of the proposed architecture.

3.4 Measurement on *PA* Silicon and *PA* Module

3.4.1 Measurement on the Assembled PA Module

The *PA* module {*Silicon+IPD*} was assembled on a *TMM10 Roger* substrate (Fig. 3.33) and characterized. Figure 3.33a represents the *silicon to IPD* interface. Figure 3.33b shows the test *PCB*. The *Chip-on-Board* assembly is protected by a resin film. The power-independent quiescent current of the driver and power stages were respectively adjusted by means of the external resistors R_{BIAS_driver} and R_{BIAS_power}. In this release, the dual adaptive bias system was integrated. For sake of safety, the filter *H*(s) was implemented as a simple 30 MHz low-pass filter and 2nd generation *LDMOS* were used.

3.4.1.1 Small-Signal Measure

The *S*-parameters of the *PA* module were characterized by means of a 4-port *Vector Network Analyzer* (*VNA*) and by selecting the differential to single mode. They are depicted in Fig. 3.34. The port 1 is coupled with the single-ended 50 Ω output port of the *PA* module. The port 2 is coupled with the differential 100 Ω input port of the *PA* module. Input return loss (S_{DD22}) is –14 dB (resp. –11 dB) at 1.75 GHz (resp. 1.95 GHz). Output return loss (S_{SS11}) is –10 dB (resp. –6 dB) at 1.75 GHz (resp. 1.95 GHz). The transfer gain (S_{SD12}) is 23 dB (resp. 18 dB) at 1.75 GHz (resp. 1.95 GHz) with a 25 dB maximum value at 1.6 GHz. The optimum band is shifted by 100 MHz which results in higher losses than the simulated results in Fig. 3.22.

This discrepancy is due to the underestimated series parasitic inductance due to inter-chip wire-bondings (see Fig. 3.33) that can be as high as 1 nH (instead of the simulated 0.6 nH inductor). Out-of-band rejection is an important feature since it relaxes the linearity level that is expected from the *PA*. Indeed, out-of-band spurious (interferers, noise…) in the neighborhood of the targeted bandwidth can generate

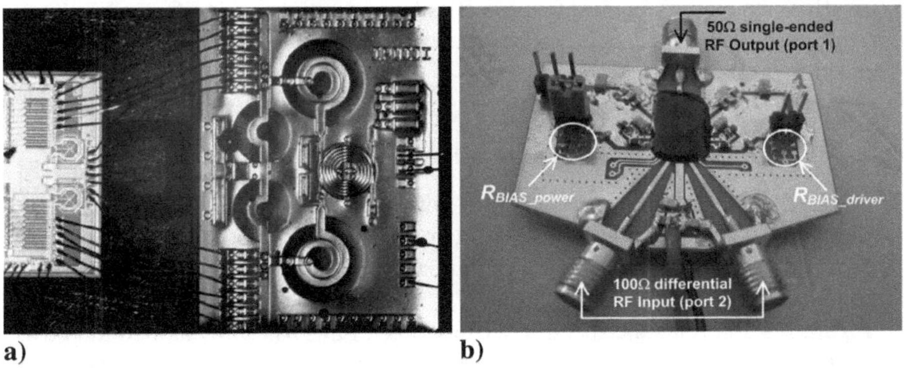

a) **b)**

Fig. 3.33 *PA* module assembly: (**a**) silicon to *IPD* interface, (**b**) *PCB* demonstrator

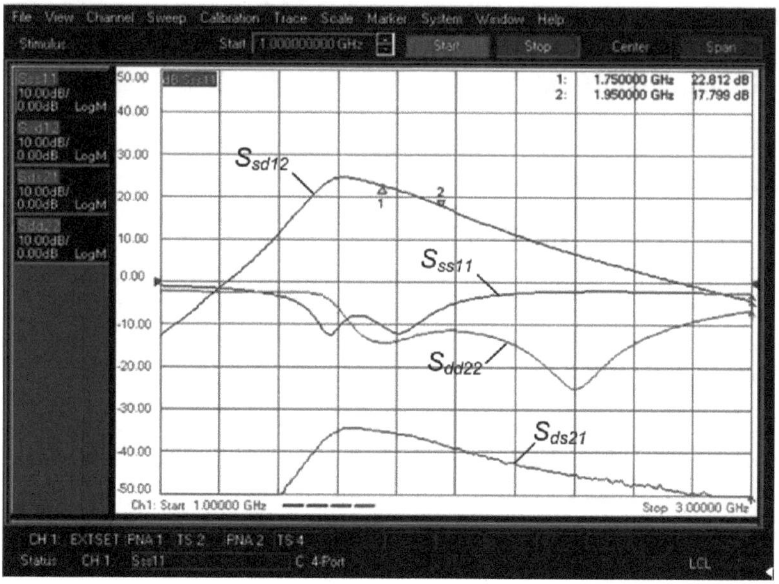

Fig. 3.34 *PA* module *S*-parameters

inter-modulation products. This is especially true in the considered architecture (Fig. 3.1) in which the ΔΣ modulator introduces substantial out-of-band noise (see further Section 3.4.1.3). The rejection performance is 33 dBc in the *GSM* band (900 MHz) and 5 dB in the 2.45 GHz *WLAN* band compared to the *WCDMA* band.

3.4.1.2 Large-Signal Measurement

The maximum output power is 27 dBm at 1.75 Gz (Fig. 3.35a) which is below the targeted 30 dBm. The power adaptive bias is depicted in Fig. 3.35b and the maximum consumed current is approximately 0.8 A which is below the expected 1.3 A. Both statements suggest that the *PA* silicon encounters significant *VSWR* and that the drain load impedance exceeds the expected 5 Ω differential load due to the inter-chip wire-bondings.

3.4.1.3 Characterization of the Overall Transceiver Demonstrator

The overall transceiver demonstrator was characterized within ST Microelectronics facilities as depicted in Fig. 3.36. This demonstrator included an *I/Q QPSK* emulator, a ΔΣ generator mother board (including a *BAW* pre-filter), a *BICMOS* power amplifier and an output *BAW* duplexer.

Figure 3.37 presents the spectral performance of this assembly at medium power level ($P_{OUT} = 13$ dBm). The specified spectral mask was fulfilled in the *GSM*-900 and *DCS*-1800 bands and especially in the *WCDMA RX* band for which spectral noise density is below the specified –126 dBm/Hz (i.e. –76 dBm(100 kHz)).

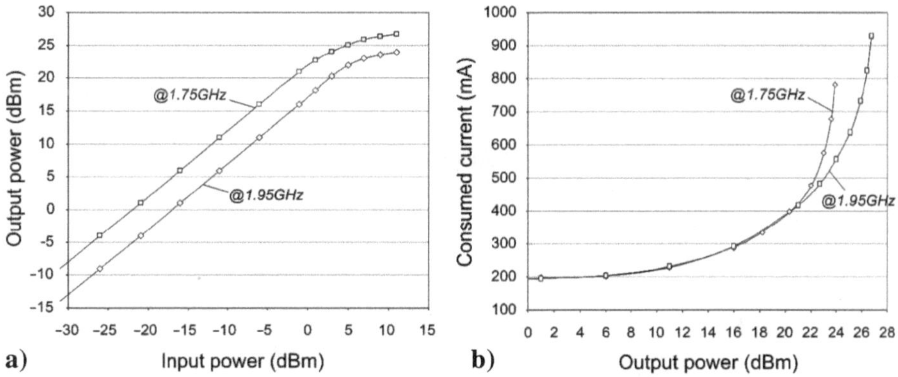

Fig. 3.35 *PA* module *CW* performance, (**a**) output power vs. input power, (**b**) current consumption vs. output power

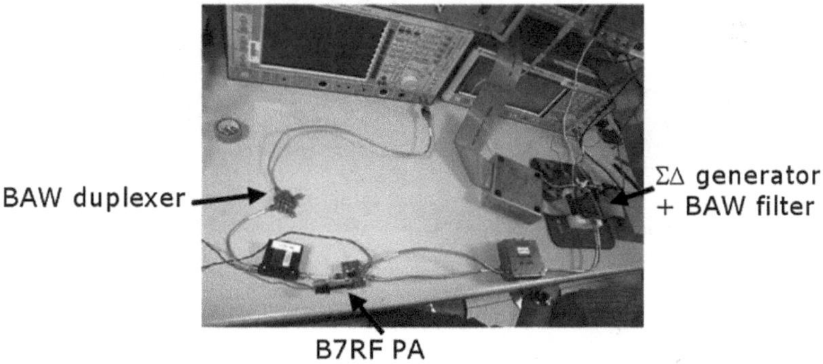

Fig. 3.36 Measurement setup for the full transmitter

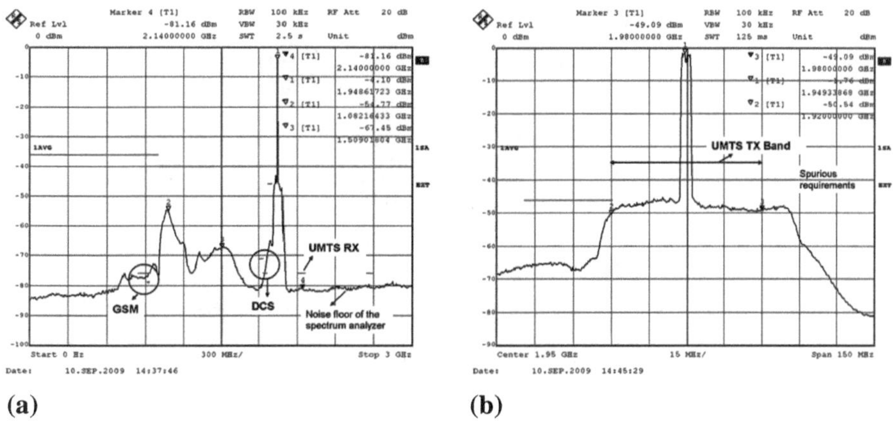

Fig. 3.37 Measured output spectrum for the full transmitter, (**a**) wide frequency span, (**b**) narrow frequency span

3.4.2 Measurement of Stand-Alone Silicon

3.4.2.1 Characterization Test-Bench

The *PA* silicon was tested stand-alone by means of on-wafer characterization. To this end, a *Süss-Microtec* single-ended load-pull test-bench was employed (Fig. 3.38) with specific calibrations at 1.75 and 1.95 GHz. The *PA* die micrograph is depicted in Fig. 3.39. The *PA* output pad is implemented according to a hybrid structure in compliance with wire-bonded and on-wafer measurement.

Figure 3.40 presents the micrograph of the power stage (a) and the driver stage (b). Some layout rules were employed to optimize the *RF* performance of

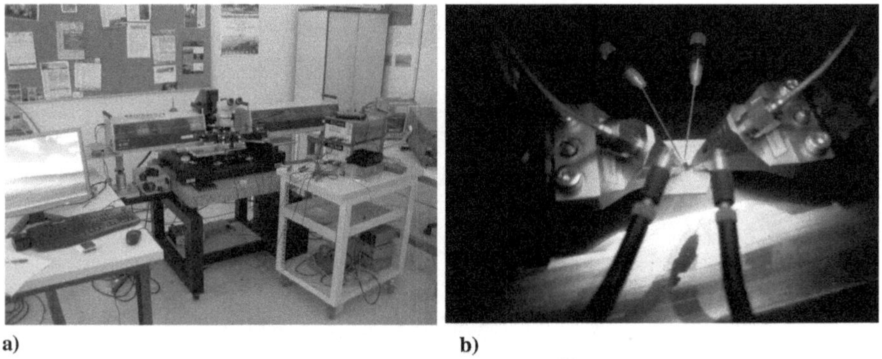

Fig. 3.38 On-wafer single-ended load-pull test bench

Fig. 3.39 Single-ended view of *PA* die micrograph

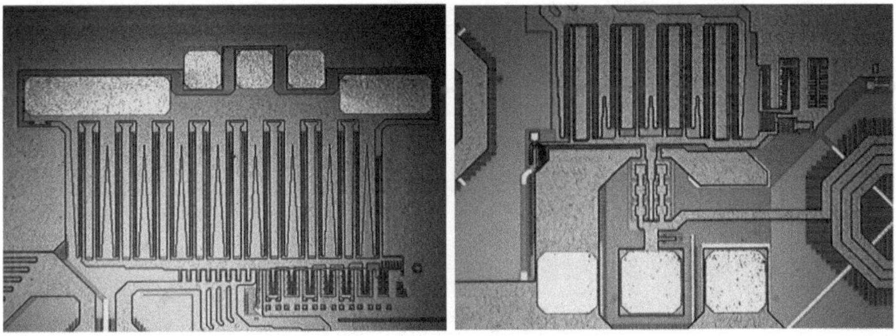

a) b)

Fig. 3.40 Zoom views of (**a**) single-ended power stage, (**b**) driver stage

the power *LDMOS*. The gate/ground capacitance was minimized by disengaging ground paths below drain pads rather than below the gate access. The drain accesses were designed as *distributed* stair-shaped paths in order ensure minimum drain capacitance while coping with electro-migration constraints.

Indeed, the maximum allowed current density is 3 mA/μm in the upper metal layer of the 0.25 μm *BICMOS* ST Microelectronics technology. The overall drain access of a power *LDMOS* that has to safely drive a current as high as 800 mA must approximately be 270 μm wide (i.e. nine 30 μm-wide fingers).

3.4.2.2 Characterization at 1.75 GHz (*DCS/EDGE* Mode)

A load-pull test was carried out and the associated power gain contours are shown in Fig. 3.41. This allowed identifying a load optimum in terms of power gain and efficiency that was found to be $\{|\Gamma_L| = 0.7;\ \angle\Gamma_L = 169°\}$, in other words $Z_{L\_opt} = (6.8 + j8)\ \Omega$. This value is limited by the test bench capabilities but remains consistent with the estimated optimum impedance in (3.32).

Consistently with the simulations, the single-ended output compression point is 27.5 dBm (Fig. 3.42) and the associated linear *PAE* is 40% (Fig. 3.43). Maximum *PAE* is 48%. Current consumption varies from 100 to 550 mA, i.e. over a 5:1 ratio. The *PAE* at the backed-off power (*OCP*1–10 dB) is as high as 10%, in contrast with the *PAE* of 5% that a class-A *PA* stage would feature in similar conditions. This clearly illustrates the benefit of the adaptive bias at medium power levels.

The output 3rd order intercept point (OIP_3) is obtained with a 200 kHz-spaced dual tone by linearly extrapolating the 1st and 3rd order harmonic waveforms. The reference point of the extrapolation is chosen as (CP_1–10 dB = 17.5 dB).

The OIP_3 is found to be approximately 50 dBm (Fig. 3.44). Such good result is due to the fact that the 3rd order inter-modulation product presents an inflexion at medium power level which is a typical consequence of adaptive bias. This was already noticed in Fig. 3.27.

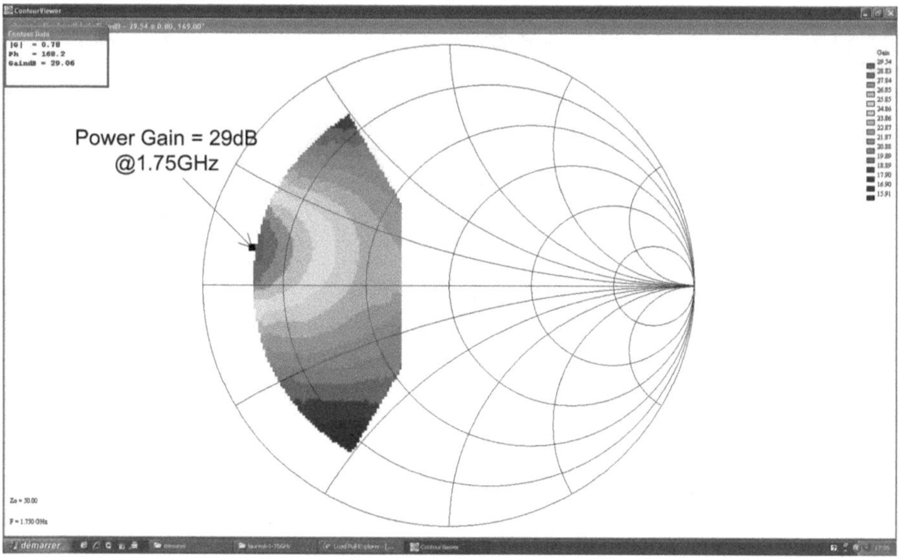

Fig. 3.41 Load-pull illustration at 1.75 GHz

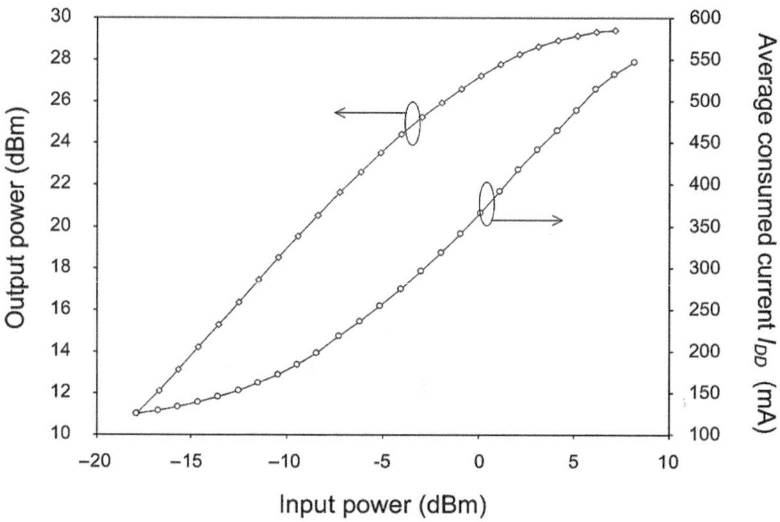

Fig. 3.42 Output power and current consumption as a function of input power

The *PA* performance was characterized with a *8PSK EDGE* modulated signal that was emulated by an Agilent ESG-4033B generator. The channel bandwidth over which the useful power is integrated is 200 kHz. Figure 3.45 presents such *EDGE* channel at the output power $P_{OUT} = 21$ dBm. The main spurious rejection profiles are depicted in Fig. 3.46. The spectral mask specification at 200 kHz is fulfilled

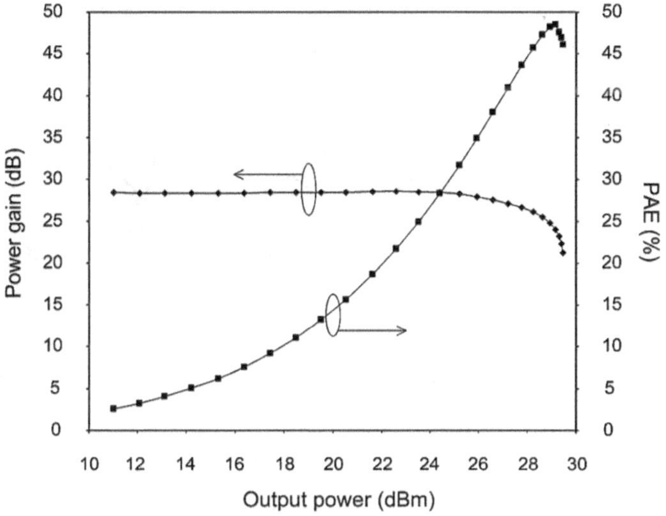

Fig. 3.43 Power gain and *PAE* as a function of output power

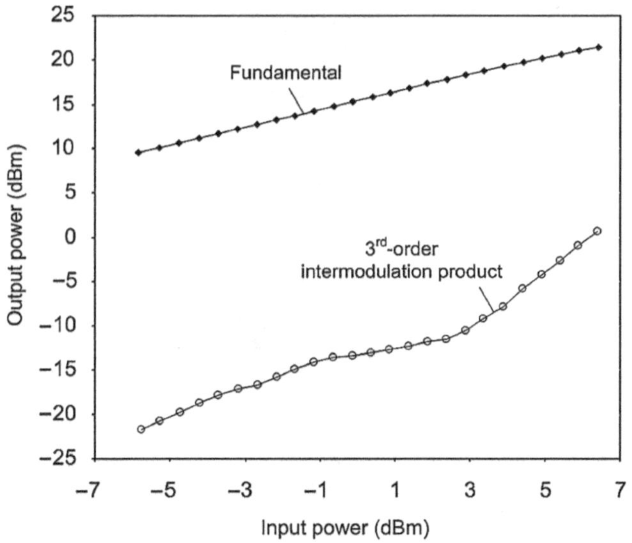

Fig. 3.44 *OIP*$_3$ illustration with a 200 kHz-dual-tone signal

until $P_{OUT} = 23$ dBm for which *PAE* is 26%. In the present case, the most stringent *EDGE* spurious rejection specification is –54 dBc@400 kHz. The maximum output power in compliance with *EDGE* spectral requirements is 21 dBm for which *PAE* is 17% (Fig. 3.46). Since *EDGE* is a phase-modulation-based standard, the distortion of its spectrum is strongly related to the performance of the *PA* in terms of *AM/PM*.

Fig. 3.45 *PA* output spectrum ($P_{OUT} = 21$ dBm) with *EDGE* signal and spectral mask specification

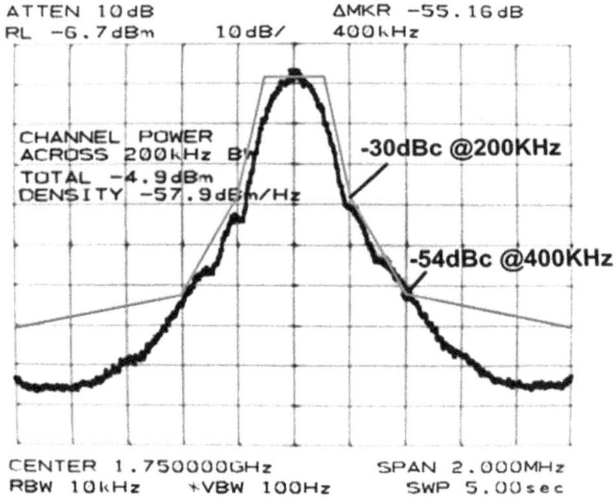

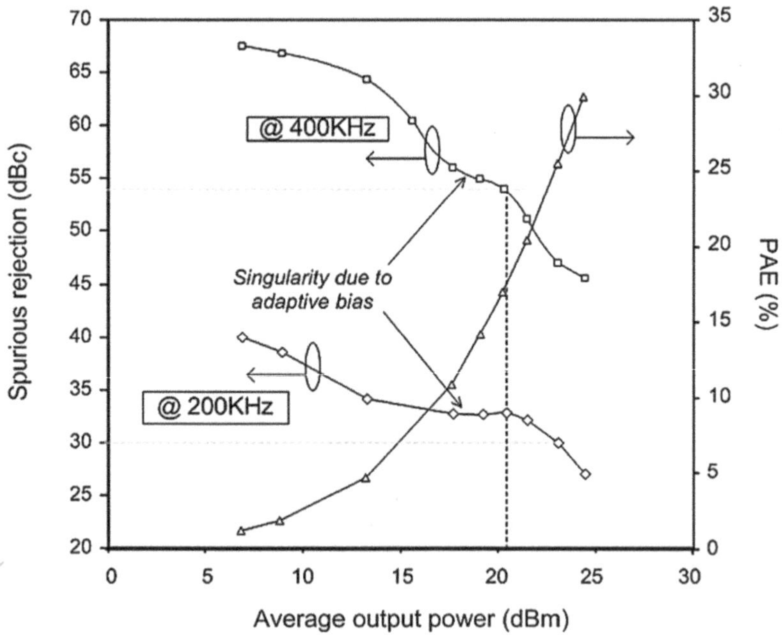

Fig. 3.46 *PA* output spurious rejection (@ 200 and 400 kHz from carrier) and the associated *PAE* with an *EDGE* modulated signal

Spurious rejection profiles in Fig. 3.46 present inflection points at medium power levels (between 15 dBm and 20 Bm). Such singularities are another typical positive consequence of adaptive bias. They allow increasing linearity and enhance the maximum affordable power. The right (upper) side of the *EDGE* channel is found to be degraded compared with the left (lower) side counterpart (Fig. 3.45) which limits the maximum average efficiency. This asymmetry highlights the impact of memory effects that arise at frequencies that are substantially far from the carrier.

3.4.2.3 Characterization at 1.95 GHz (*WCDMA* Mode)

The single-ended output compression point is up to 27.5 dBm at 1.95 GHz (Fig. 3.47). The associated current consumption and *PAE* are respectively 410 mA (Fig. 3.47) and 51% (Fig. 3.48). Maximum *PAE* is 57% and *PAE* @(OCP_1 – 10 dB) remains good, up to 12% which again demonstrates the interest of this architecture (Fig. 3.48). The comparison of measurement and simulation presents good agreement. Meanwhile, the measured current consumption is found to be slightly decreased whereas the measured power gain presents flatter response and does not feature the typical modulation of adaptive bias system. This is essentially due to temperature increase whose impact is hardly taken into account in simulations. A slight discrepancy between measure and simulation lies in phase modulation (*AM*/*PM*).

AM/*PM* is found to be –14° instead of –8° at $P_{OUT} = 28$ dBm. This is consistent with the spectral mask degradation that was observed with a *8PSK* modulated signal (see Section 3.4.2.2).

It is worth noting that, due to the relatively high load impedance (singled-ended 6 Ω load), both simulated and measured *AM*/*PM* waveforms are monotonous (no singularity) in contrast to what was observed in low impedance conditions (i.e. differential 6 Ω load in Fig. 3.25).

The output 3rd order intercept point (OIP_3) is obtained in a similar manner to Section 3.4.2.2, *using* (OCP_1 – 10 dB = 17.5 dB) as a reference point and for several spacing values.

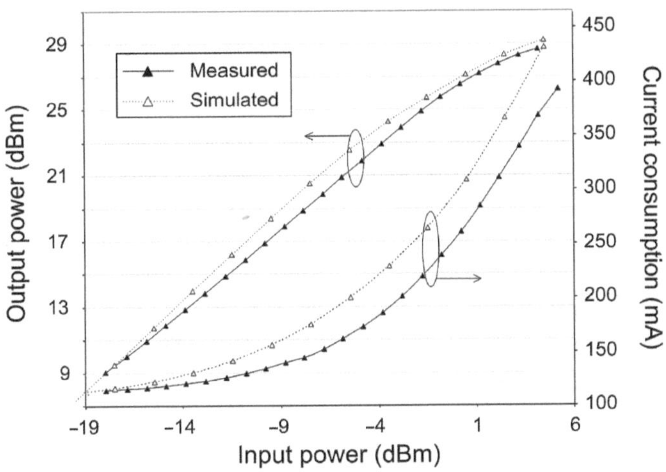

Fig. 3.47 Measured and simulated output power and current consumption vs. input power

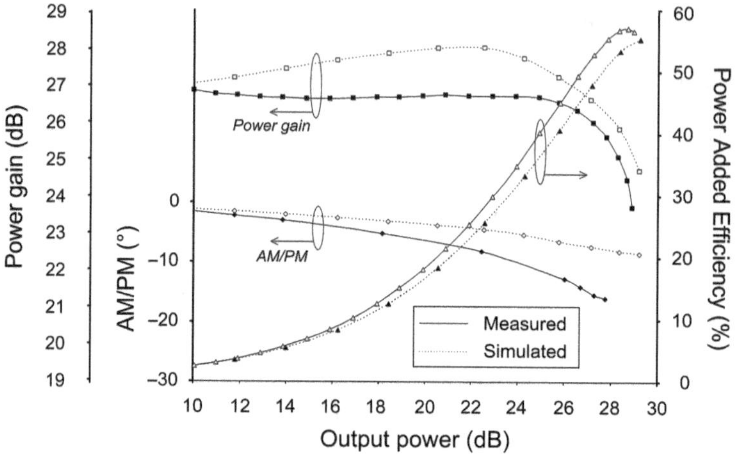

Fig. 3.48 Measured and simulated power gain, *AM/PM* and *PAE* vs. of output power

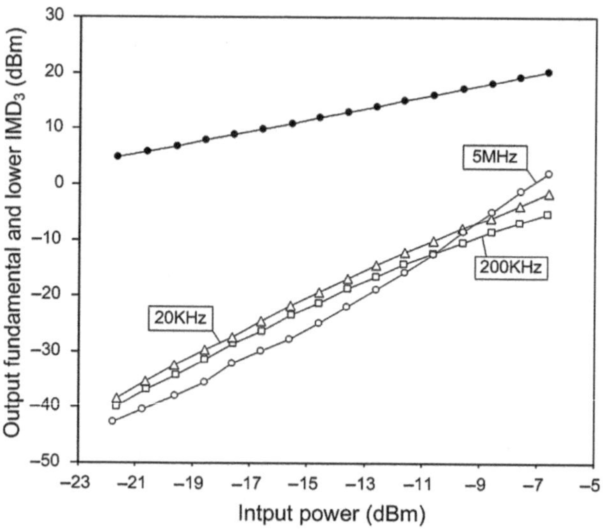

Fig. 3.49 OIP_3 illustration with a 5 MHz-dual-tone signal

The OIP_3 is found to be nearly infinite with a 200 kHz spacing whereas it is only 30 dBm with a 5 MHz spacing (Fig. 3.49). Such frequency dependency highlights the memory effects that are involved in the *PA* behavior. A discussion on their physical and electrical explanations will be provided in Section 3.4.3.

The *PA* performance was characterized with a *HPSK* modulated signal featuring a 3.5 dB crest factor (Agilent E4433B generator). In order to highlight the memory effects, measurements were carried out for various chip rate values (see snapshots in Fig. 3.50a–c). A reversed image of the envelope was monitored and the adaptive bias bandwidth was characterized by probing the voltage at the test point I_{Q_bias} (see Fig. 3.5). Figure 3.50d presents the image of a *HPSK* modulated envelope.

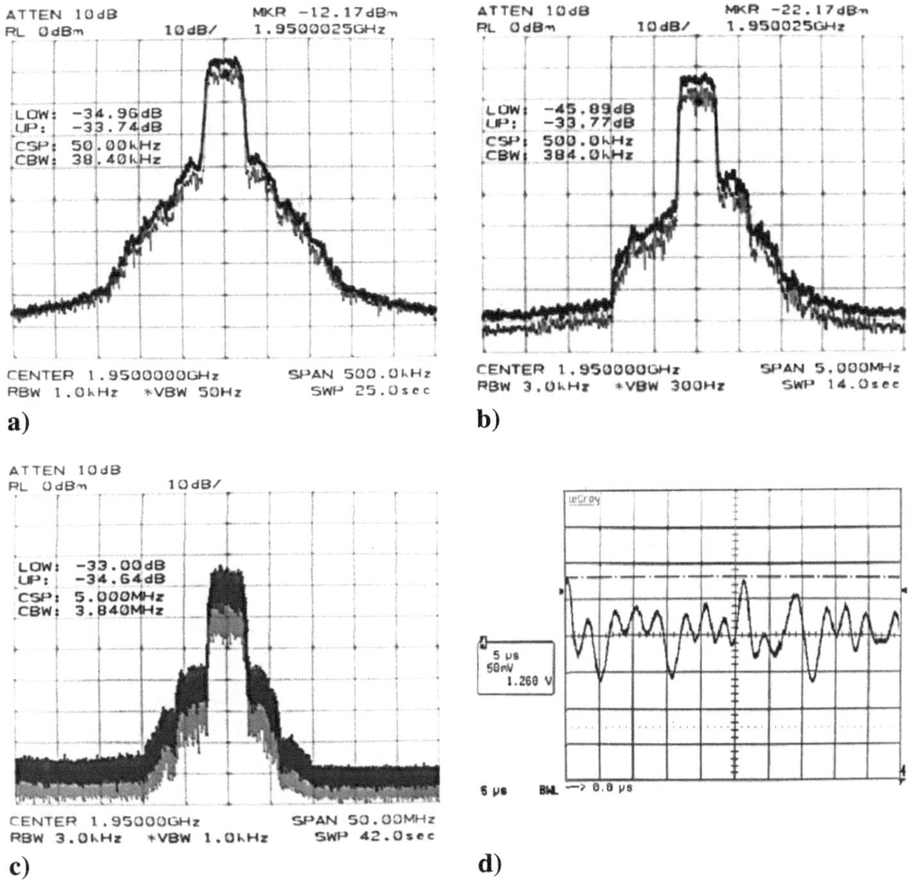

Fig. 3.50 Output spectrum for various chip rate values, (**a**) 38.4 kcps, (**b**) 384 kcps, (**c**) 3.84 Mcps (*WCDMA*), (**d**) image of reversed *HPSK* envelope (probed at mixer output)

The linearity/efficiency performance is summarized in Fig. 3.51 as a function of chip rate. The maximum affordable output power (at $ACLR = 33$ dBc) can be as high as 26.5 dBm for very slow chip rates. The proposed adaptive bias architecture reduces the necessary power back-off to cope with *HPSK* modulation down to only 1 dB (instead of 3 dB as generally assumed for a *WCDMA* class-A *PA*). The associated average *PAE* is up to 42% which is substantially higher than the theoretical 25% *PAE* (50%/2) of a 3 dB-backed-off *PA*. For faster chip rates, the linearity/efficiency decreases due to complex non-modelled memory effects, to the bandwidth limitation of the adaptive bias system and to unexpected out-of-band terminations.

This is particularly visible in Fig. 3.50b where strong imbalance is observed between lower and upper adjacent channels (–46 dBc vs. –33 dBc of *ACLR*). At 3.84 Mcps (*WCDMA* specification), the maximum affordable power is 21 dBm and its associated average *PAE* is 22%.

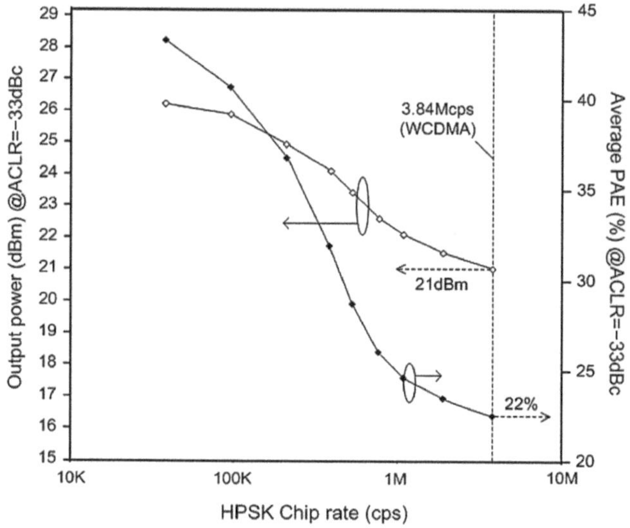

Fig. 3.51 *ACLR* and average *PAE* for various chip rate values

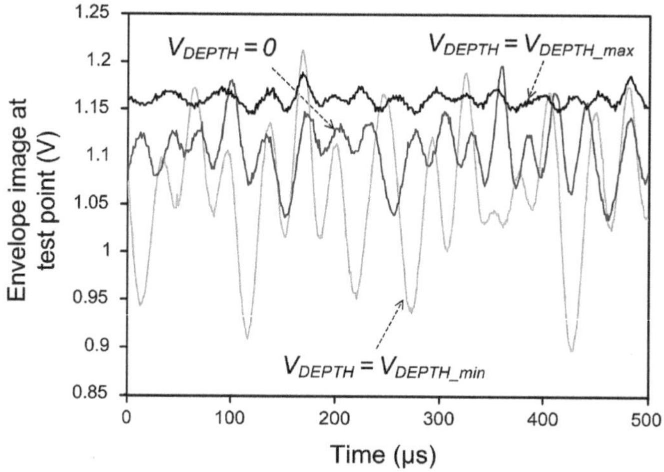

Fig. 3.52 Measured reversed image of *HPSK* envelope (probed at test point/mixer output)

3.4.2.4 Impact of Finely-Tuned Adaptive Bias on Linearity

Figure 3.52 illustrates to what extent the reversed envelope image (probed at the test point) is modulated by V_{DEPTH} in time domain (38.4 kcps *HPSK* modulated signal).

Figures 3.53 and 3.54 depict the impact of V_{DEPTH} tuning respectively on IMD_3 and *ACLR* (worst case). The effect of V_{DEPTH} is mitigated for spacing and chip rates above 1 MHz due to excess loop time response that is insufficiently compensated by

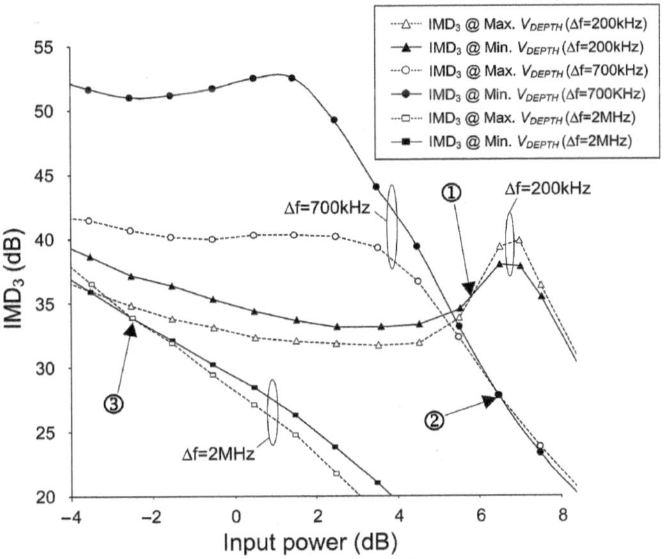

Fig. 3.53 *IMD_3* power amplifier response for various dual-tone spacing values and V_{DEPTH} values

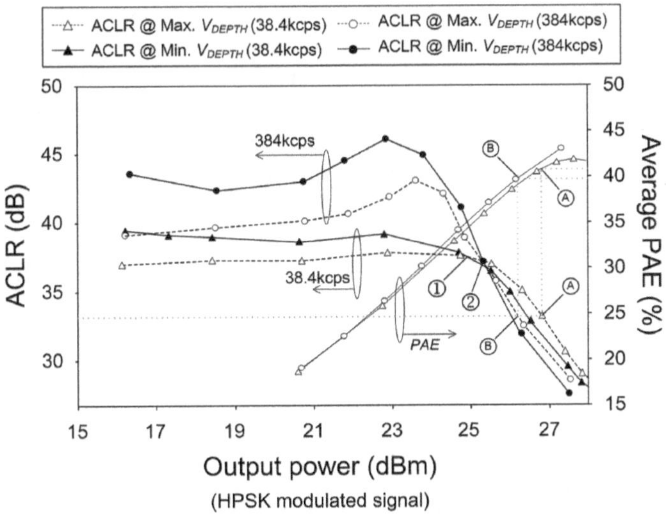

Fig. 3.54 *ACLR* power amplifier response for various chip rate and V_{DEPTH} values

$A(\omega)$. The control range of IMD_3 can be up to 10 dB (for a 700 kHz spacing) at low power level. At higher power levels, the reversal of V_{DEPTH} polarity (depicted by the points ①,②, and ③) allows linearity to be enhanced whereby IMD_3 is increased (e.g. by 2 dB for a 200 kHz spacing). Identically to what is stated with IMD_3, *ACLR* can be improved by correctly selecting the polarity and magnitude of V_{DEPTH} voltage.

For example a 2 dB (resp. 5 dB) *ACLR* improvement is obtained at low power levels with a 38.4 kcps chip rate (resp. a 384 kcps chip rate) if V_{DEPTH} is set to its lower value. Above a reversal level (represented by ①, and ② marks), *ACLR* is optimized by setting V_{DEPTH} to its highest value. The points A, and B depict the maximum linear power (for which *ACLR* is 33 dBc) and their associated *PAE* for various chip rate values.

These measures demonstrate that the fine tuning of the envelope injection makes linearity optimization possible over specific power ranges at the cost of no current consumption (assuming the mixer consumption can be neglected). Reducing V_{DEPTH} at low power level allows preventing over-injection while increasing V_{DEPTH} at high power level allows extending the *PA* power capability.

3.4.2.5 Robustness to *VSWR*

Ruggedness to *VSWR* variations is an important feature from the point of view of reliability and determines if an external isolator is required or not. Indeed, removing the isolator from the overall front-end reduces insertion losses and the *Bill-Of-Material* (i.e. cost and area). On the one hand, the output *IPD* network provides substantial isolation. Indeed, the *IPD* network converts a 10:1 *VSWR* locus at antenna into a 2:1 *VSWR* locus at the *PA* silicon output. On the other hand, it can be stated from Fig. 3.55 that the more output load increases, the lower the current variation with respect to power. This is consistent with what was expected from theory and simulations. If the output load increases, both the power stage input impedance and the gate voltage magnitude decrease (the former due to Miller effect). Consequently, current consumption is expected to decrease, as shown in (3.17). As in Section 2.4.3.6, such current *curbing* plays an useful protection role

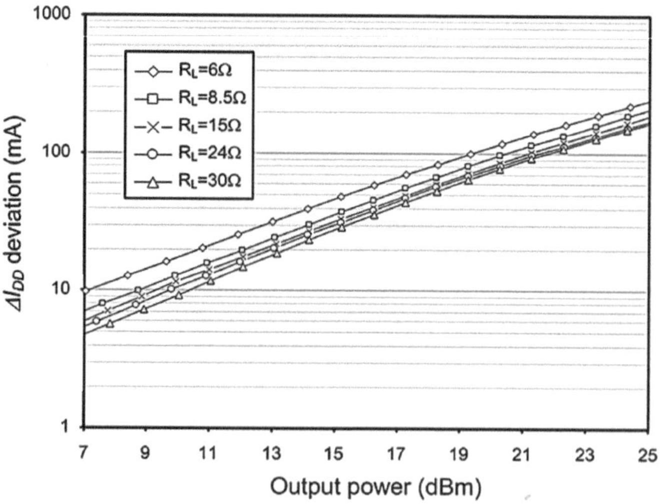

Fig. 3.55 Measured adaptive bias sensitivity to *VSWR* mismatch

since it results in reduced drain over-voltage. The derivative of the drain current with respect to the input power (∂I_{DD} / ∂P_{IN}) is measured to be 0.465 A/W in nominal conditions, whereas the second derivative with respect to the output load ($\partial^2 I_{DD}$ / ∂P_{IN} ∂R_L) is limited to 24 mA/W/Ω. Figure 3.55 demonstrates that the proposed architecture allows the power amplifier to withstand an output power as high as 25 dBm in the condition of $VSWR = 10{:}1$ ($R_L = 30\ \Omega$).

3.4.3 Discussion and Conclusion

The *PA* performances for various standards (*DCS*, *EDGE*, *WCDMA*) are summarized in Table 3.1. All values refer to stand-alone silicon single-ended characterization, in the nominal load condition. The maximum power added efficiency is obtained in Continuous-Wave mode. The dual-tone spacing is 200 kHz for *DCS* and *EDGE* and 5 MHz for *WCDMA*. The nominal output power is defined as: the maximum power in Continuous-Wave mode for *DCS*, the maximum power with a *8PSK* modulated signal for which the spurious rejection at 400 kHz is –54 dBc for *EDGE*, the maximum power with a *HPSK* modulated signal for which *ACLR* is 33 dBc for *WCDMA*.

The continuous-wave power performances of the proposed architecture are comparable with state-of-the-art literature (Table 3.2) at the expense of low die area (1×1.2 mm^2), low consumption, low complexity, and low *Bill-of-Material*.

The adaptive bias properly operates over a 4:1 power range (see current consumption waveform in Fig. 3.47) and reduces the impact of crest factor on efficiency

Table 3.1 *PA* performance summary

		DCS	*EDGE*	*WCDMA*
Technology	–	*BICMOS SiGe* 0.25 μm		
Operating frequency range	(MHz)	1710–1785		1920–1980
Nominal output power (*RMS*)	(dBm)	29	21	21
Minimum linear gain	(dB)	28	28	26
Max. power added efficiency	(%)	47		57
Output 3rd order intercept point	(dBm)	50 (200 kHz pacing)		30 (5 MHz spacing)
Power supply	(V)	3.5		

Table 3.2 Performance comparison with literature

	PAE @OCP_1	*PAE* @(OCP_1 –10 dB)	*PAE* max
Apel et al. (2008)	31% @28 dBm	8% @18 dBm	–
Bakalski et al. (2008) (low power mode)	33% @28 dBm	9% @18 dBm	47%
Presti et al. (2008)	45% @25 dBm	15% @15 dBm	50%
Pinon et al. (2008)	50% @27 dBm	15% @17 dBm	–
This work	51% @27.5 dBm	12% @17.5 dBm	57%

since the necessary back-off is reduced from 3.5 to 1 dB (Figs. 3.47 and 3.51). The possibility to modulate the injected envelope and to improve linearity over the power range is demonstrated by setting the V_{DEPTH} voltage to its optimum value accordingly (Figs. 3.53 and 3.54).

The characterization with *EDGE* or *WCDMA* modulated signals has enlightened typical limitations of adaptive bias power amplifiers, namely the bandwidth over which the adaptive bias system properly operates and the control of *AM/PM*. Indeed, adaptive bias hardly tracks fast envelope variations, thus increasing *ACLR* and reducing the maximum linear affordable power. These issues are all the more critical as channel bandwidths are getting wider up to 20 MHz for the *Long-Term-Evolution*. In order to overcome the discrepancies due to high channel bandwidths, three factors must be taken into account in the preliminary steps of the design flow. Firstly, thermal memory effects play an important role in the observed increase in *AM/PM*. Indeed, temperature variations due to the envelope swing dynamically modulate the *RF* properties of the *PA* and hence introduce a frequency-dependent feedback. Depending on the polarity of thermal dependencies ($\partial\ g_{m,1}/\partial\ T°$, $\partial\ C_{gs,1}/\partial\ T°$, amongst others), phase delay or phase advance is introduced in the *PA* power response, and the non-linear ε vector in equation (3.24) is shifted. Such phenomenon can be compensated to some extent by properly modelling the equivalent thermal network of the power device and by optimizing the frequency behavior of the base-band filter *H(f)*. Second, special care must be paid to the modelling of power devices' non linearities, especially at low bias levels. The capacitances of *LDMOS* devices proposed in *B7RF* technology (ST Microelectronics *BICMOS* 0.25 μm) have been characterized (Canepari et al. 2005), and some inaccuracies have been reported compared to the actually employed models. Indeed, unlike *MOS* devices, *LDMOS* present a distributed drain, whereby capacitances feature sharp variations with respect to the bias level. Singularities are observed close to threshold bias level and the non-linear capacitance $C_{gs,2}$ is found to become alternately positive and negative at low bias level. Therefore, it is of no surprise that a negative phase shift is observed in ε (ω_{RF} , $\Delta\omega-\omega_{RF}$) according to (3.11), (3.16), and (3.24), even for low $\Delta\omega$ spacings, whereby an increased imbalance is encountered between lower and upper adjacent channels (Figs. 3.45 and 3.50b). Identically, inaccuracies in the modelling of 3rd order input capacitances $C_{gs,3}$ also result in an underestimation of the actual *AM/PM* (Fig. 3.48). Lastly, the out-of-band terminations at twice the carrier frequency $2\omega_{RF}$ and at the base-band frequency $\Delta\omega$ play a role in the 2nd order voltage gain $A_{V,2}$ of the power stage. Consequently, the response of the power adaptability factor is impacted according to (3.5) and (3.11). Out-of-band terminations must be considered during the *PA* design (see Section 3.2.1, dedicated to the *IPD* passive network) or during the *PA* load-pull characterization by optimizing the output load at $2\omega_{RF}$. The output load at $\Delta\omega$ is generally poorly controlled and may not be neglected if the cables used in the test bench are chosen long with respect to the wavelength at $\Delta\omega$. Providing the thermal/electrical second order parameters of a *PA* are accurately modelled, the architectures proposed in this chapter allows efficiency to be improved while leaving linearity unchanged over broad channel bandwidths.

References

Aikio JP, Rahkonen T (2005) Detailed distortion analysis technique based on simulated large-signal voltage and current spectra. Trans Microw Theory Tech Atlanta, USA. 53(10):3057–3066. doi:10.1109/TMTT.2005.855132

Apel T, Henderson T, Tang YL, Berger O (2008) Efficient three-state WCDMA PA integrated with high-performance BiHEMT HBT/E–D pHEMT process. Proc Radio Freq Integr Circuits Symp, pp 149–152. doi:10.1109/RFIC.2008.4561406

Bakalski W, Zannoth M, Asam M, Thomann W, Kapfelsperger B, Pfann P, Berkner J, Hepp C, Steltenpohl A, Österreicher W, Rampf E (2008) A load-insensitive quad-band GSM/EDGE SiGeC-bipolar power amplifier with a highly efficient low power mode. Proc Radio Week Symp, Orlando, USA. pp 203–206. doi:10.1109/RWS.2008.4463464

Boumaiza S, Ghannouchi FM (2003) Thermal memory effects modelling and compensation in RF power amplifiers and predistortion linearizers. Trans Microw Theory Tech 51(12):2427–2433. doi:10.1109/TMTT.2003.820157

Brinkhoff J (2004) Bandwidth-dependent intermodulation distortion in FET amplifiers. PhD Thesis, Macquarie University, Sydney, NSW, Australia

Canepari A, Bertrand G, Giry A, Minondo M, Blanchet F, Jaouen H, Reynard B, Jourdan N, Chante JP (2005) LDMOS modelling for analog and RF circuit design. Proc Eur Solid-State Device Res Conf, Grenoble, France. pp 469–472. doi:10.1109/ESSDER.2005.1546686

Cha J, Yang Y, Shin B, Kim B (2003) An adaptive bias controlled power amplifier with a load-modulated combining scheme for high efficiency and linearity. Microw Theory Tech Symp Dig, pp 81–84

Deng J (2005) High-efficiency and high-linearity SiGe BiCMOS power amplifiers for WCDMA handset applications. PhD Thesis, University of California, San Diego, CA, USA

Flament A, Giraud S, Bila S, Chatras M, Frappe A, Stefanelli B, Kaiser A, Cathelin A (2009) Complete BAW filtered CMOS 90 nm digital RF signal generator. Proc Int Northeast Workshop Circuits Syst Conf, Toulouse, France. pp 1–4. doi:10.1109/NEWCAS.2009.5290423

Kim HT, Lee KH, Choi HK, Choi JY, Lee KH, Ryu JP GH, Jeon YJ, Han CS, Kim K, Lee K (2004) High efficiency and linear dual chain power amplifier without/with automatic bias current control for CDMA handset applications. Proc Eur Microw Conf, Amsterdam, The Netherlands. pp 337–340

Kimura K, Seki M, Matsumura N, Honjo K (2004) Improvement in ACLR asymmetry for W-CDMA InGaP/GaAs HBT power amplifier. Proc Eur Microw Conf, Amsterdam, The Netherlands. pp 333–336

Leung VW, Deng J, Gudem PS, Larson LE (2005) Analysis of envelope signal injection for improvement of RF amplifier intermodulation distortion. J Solid-State Circuits 40(9): 1888–1894. doi:10.1109/JSSC.2005.848176

Leyssenne L, Jarry P, Pham JM, Kerhervé E, Saias D (2006) Analysis of temperature modulation on a SiGe power amplifier non linearity. Proc Bipolar/BiCMOS Technol Meet, Maastricht, The Netherlands. pp 1–4. doi:10.1109/BIPOL.2006.311130

Leyssenne L, Kerhervé E, Deval Y, Belot D (2007) Procédé et dispositif d'amplification de puissance d'un signal radiofréquence. Patent N°B09-1273FR FZ/CRA/AE, France

Liu T, Boumaiza S, Ghannouchi FM (2005) Deembedding static nonlinearities and accurately identifying and modelling memory effects in wide-band RF transmitters. Trans Microw Theory Tech 53(11):3578–3587. doi:10.1109/TMTT.2005.857105

Maas S (2003) Nonlinear microwave and RF circuits. Artech House, Norwood

Mizusawa N, Kusunoki S (2005) Third- and fifth-order baseband component injection for linearization of the power amplifier in a cellular phone. Trans Microw Theory Tech. doi:10.1109/MWSYM.2005.1516997

MOBILIS FP6 IST Project (2006) Mixed SiP and SoC Integration of power BAW filters for digital wireless transmissions. IST specific targeted research or innovation project No 027003. http://www.ist-mobilis.org Accessed 15 Dec 2010

Noh YS, Park CS (2004) An intelligent power amplifier MMIC using a new adaptive bias control circuit for W-CDMA applications. J Solid-State Circuits 39(6):967–970. doi:10.1109/JSSC.2004.827804

Oka T, Hasegawa M, Fujita K, Yamashita M, Hirata M, Kawamura H, Sakuno K (2005) Enhanced linearity and efficiency of HBT power amplifiers for 5-GHz wireless-LANs. Microw Theory Tech Symp Dig, Long Beach, USA. doi:10.1109/MWSYM.2005.1516689

Paulin R, Garcia P, Mouis M, Belot D (2004) Study of compensations between 2nd and 3rd order intermodulations in a 0.25 μm-BiCMOS single SiGeC-HBT amplifier for IIP3 optimization. Asia-Pac Microw Conf Delhi, India.

Pinon V, Hasbani F, Giry A, Pache D, Garnier C (2008) A single-chip WCDMA envelope reconstruction LDMOS PA with 130 MHz switched-mode power supply. Proc Int Solid-State Circuits Conf, San Francisco, USA. pp 564–636. doi:10.1109/ISSCC.2008.4523308

Presti CD, Carrara F, Palmisano G (2008) A high-resolution 24–dBm digitally-controlled CMOS PA for multi-standard RF polar transmitters. Proc Eur Solid-State Circuits Conf, Amsterdam, The Netherlands. pp 482–485. doi:10.1109/ESSCIRC.2008.4681897

Sahu B (2004) An integrated dynamically adaptive energy management framework for linear RF power amplifiers. GEDC Ind Advis Board, Atlanta, USA. April 2004

Shirakawa AA, Pham JM, Jarry P, Kerherve E, Dumont, F, David JB, Cathelin A (2006) A high isolation and high selectivity ladder-lattice BAW-SMR filter. Proc Eur Microw Conf, Manchester, UK. pp 905–908. doi:10.1109/EUMC.2006.281067

Tanoue T, Ohnishi M, Matsumoto H (2005) Switch-less-impedance-matching type W-CDMA power amplifier with improved efficiency and linearity under low power operation. Microw Theory Tech Symp Dig, Long Beach, USA. doi:10.1109/MWSYM.2005.1516693

Vuolevi J (2001) Analysis, measurement and cancellation of the bandwidth and amplitude dependence of intermodulation distortion in RF power amplifiers. PhD Thesis, University of Oulu, Finland

Wang C (2003) CMOS power amplifiers for wireless communications. PhD Thesis, University of California, San Diego, CA, USA

General Conclusion

This book has explored two families of adaptive bias power amplifiers, i.e. discretized switched-cells power amplifiers and adaptive gate bias power amplifiers have been investigated. The former family implies the fundamental issue of quantization noise that reduces *Signal-to-Noise/Distortion-Ratio*. Two architectures circumventing the noise issue have been derived on this principle. They have been investigated at mixed transistor/system level. Even though they behave differently, they are both ruled by the same trade-offs and require a proper analysis of the appropriate resolution/over-sampling ratio.

An open-loop topology including a *Built-In-Current-Sensor* and a closed-loop topology were proposed. The first one detects the input envelope and dynamically reconfigures the discretized power stage adaptively. It presents good results and can be finely calibrated by means of external currents. Nevertheless, it presents reduced *EVM* performance at low power levels. The second closed-loop topology solves this issue. It fundamentally tracks the instantaneous *EVM* variations and dynamically reconfigures the *PA* core. Both discretized topologies are limited by the fact that they do not cope with *AM/PM*. Another limitation is the out-of-band noise level that makes those architectures better candidates for *TDD* applications than for *FDD* applications such as *WCDMA*. In Chapter 3, a more classic current envelope tracking was investigated. Its fundamental novelty lies in the second-order accurate envelope injection system that allows optimizing linearity at any power level and circumventing the linearity/efficiency trade-off.

Two demonstrators were realized on silicon with ST Microelectronics 0.25 μm *BICMOS* technology. A *HBT* bypass-based differential power amplifier was developed for *WCDMA* application in the frame of *Asturies* project. Despite the non-ideal behaviour of the output ring balun, it features a 23.3 dBm output compression point in high power mode and a 6.6 dBm output compression point in bypass mode. It also features a 20.3 dB power gain in high power mode and a 0.4 dBm output compression point in bypass mode. By bypassing the power stage, the current consumption drops from 550 to 50 mA.

In the frame of *MOBILIS* project, a low die area dual adaptive bias differential power amplifier dedicated to *DCS/EDGE/WCDMA* was realized. In single-ended mode, its power gain is 26.5 dB at 1.95 GHz, its output compression is characterized

L. Leyssenne et al., *Reconfigurable RF Power Amplifiers on Silicon for Wireless Handsets*, Analog Circuits and Signal Processing, DOI 10.1007/978-94-007-0425-1, to be 27.5 dBm. The maximum power added efficiency is 57%. Its *PAE* is 51% at *OCP*1 and 12% at (*OCP*1-10 dB). This makes the proposed topology nearly comparable with state-of-the-art integrated *EER* architectures at the cost of reduced die area and *Bill-of-Material* (since no external choke is required). The PA response to narrow channel-bandwidth modulated signal remains efficient and consistent with simulations. This was demonstrated with a 200 kHz-wide *EDGE* channel and low-chip-rate *HPSK* modulated signal. In this latter case, the maximum allowed output power (for which *ACLR* = 33 dBc) can be up to 26.5 dBm with a 42% *PAE*, which represents a 1 dB back-off only (with regard to OCP_1). This is substantially lower than the 3.5 dB back-off (= *Peak-to-average ratio*) that was originally expected. When the envelope swing is too fast, the adaptive bias does not behave as efficiently and memory effects are found to degrade linearity. This analysis stresses the importance of thermal and electrical memory effects on the linearity performance of an adaptive bias power amplifier. Indeed, they are generally inaccurately taken into account in simulations. For example, dynamic temperature variations are poorly handled by simulators. This also raises the issue of the accurate modelling of electrical (non-linear) *LDMOS* parameters at low bias level especially close to threshold level.

Several future prospects to this work can be proposed. First, based on the discrete architectures that were presented in Chapter 2, a combined adaptive/Cartesian could be investigated so as to take phase distortion into account and better control linearity. Moreover, the digital output control word that is supplied by the power amplifier could also be exploited by an integrated Digital Signal Processing either to better calibrate the power amplifier parameters or to implement an internal digital predistortion. Another point is to investigate the integration of a digital decimator filter to reduce the out-of-band noise level. Second, based on the continuous adaptive bias that was presented in Chapter 3, further investigation efforts may be put on the analysis of memory effects and the advantage that could be drawn from the baseband filters $H(s)$ and $A(s)$ in order to reduce/magnify spectral regrowth asymmetry. Moreover, the integration of thermal sensors could also be analyzed. This way, it may become possible to monitor the dynamic variations of temperature as well as extract the *PA* effective thermal impedance.

Appendix A
Impact of Base/Emitter Degeneration on *HBT* Self-Heating Behavior

Table A.1 Maple script dedicated to the computation of a degenerated *HBT* collector current (including self-heating) as a function of the collector voltage V_{CC} for a fixed degenerated base voltage V_{IN}

```
>T0:=298;
>k:=1.38*10^(-23);
>q:=1.6*10^(-19);
>vcc:=0.8;
 /// Input degenerated base voltage
>vin:=0.8;
>vearly:=250;
 /// Number of parallel HBT
>mult:=1000;
 /// Emitter resistor degeneration
>re:=0.1;
 /// Base resistor degeneration
>rb:=0;
>rl:=0.0;
 /// Thermal resistance
>rth:=30;
>is00:=1.0*10^(-11);
>xti:=3;
>xtb:=-1.5;
>nc:=1.0;
>beta0:=200;
>eg0:=0.875;
>alfateg:=0.003;
>betateg:=600;
 /// Relative temperature deviation
>tdev:=i->1+(i*rth*(vcc-(rl+re)*i))/T0;
>is0:=i->is00*(tdev(i)^xti);
>eg:=i->eg0-(alfateg*(T0*tdev(i))^2)/
 (betateg+(T0*tdev(i)));
⇒
>beta:=i->beta0*(tdev(i)^xtb);
>ut:=i->k/q*T0*tdev(i);
>eth:=i->k*T0*tdev(i);
>ic:=i->mult*is0(i)/(1-(vcc-rl*i)/vearly)
*exp(vin/(nc*ut(i))-
i*(re+rb/beta(i))/(nc*ut(i))-eg(i)/ut(i));
..../// Routine beginning
>vcc:=0.8;
vin:=1.0;
fy:=fopen(icc,WRITE,TEXT);
fw:=fopen(temp,WRITE,TEXT);
fz:=fopen(vccc,WRITE,TEXT);
 /// Collector voltage sweep
while vcc<5.8 do
 /// Computation of collector current
y0:=fsolve(ic(i)-i=0);
w0:=eval(tdev(i),i=y0);
Y0:=array([y0]);
 /// Temperature computation in °C
W0:=array([w0*T0-273]);
Z0:=array([vcc]);
writedata(fy,Y0,float);
writedata(fw,W0,float);
writedata(fz,Z0,float);
vcc:=vcc+0.1;
end do;
fclose(fy);
fclose(fw);
fclose(fz);
```

Table A.2 Maple script dedicated to the computation of a degenerated *HBT* collector current (including self-heating) as a function of the degenerated base voltage V_{IN} for a fixed current bias I_{IN} and a fixed collector voltage V_{CC}

```
>T0:=298;
>k:=1.38*10^(-23);
>q:=1.6*10^(-19);
>vcc:=3.3;
  /// Input base bias current
>Iin:=0.005;
  /// Number of parallel HBT
>mult:=1000;
  /// Emitter resistor degeneration
>re:=0.1;
  /// Base resistor degeneration
>rb:=0;
>rl:=0;
  /// Thermal resistance
>rth:=60;
>is00:=1.0*10^(-11);
>xti:=3;
>xtb:=-1.5;
>nc:=1.0;
>beta0:=200;
>eg0:=0.875;
>alfateg:=0.003;
>betateg:=600;
  /// Relative temperature deviation
>tdev:=i->1+(i*rth*(vcc-(rl+re)*i))/T0;
>is0:=i->is00*(tdev(i)^xti);
>eg:=i->eg0-(alfateg*(T0*tdev(i))^2)/
  (betateg+(T0*tdev(i)));
>beta:=i->beta0*(tdev(i)^xtb);
>ut:=i->k/q*T0*tdev(i);
⇒
>eth:=i->k*T0*tdev(i);
>ic:=i->Iin*beta(i);

..../// Routine beginning
>vcc:=3.3;
Iin:=0.0000001;
fy:=fopen(icc,WRITE,TEXT);
fw:=fopen(vin,WRITE,TEXT);
fz:=fopen(temp,WRITE,TEXT);
  /// Base current sweep
while Iin<0.010 do
  /// Collector current computation
y0:=fsolve(ic(i)-i=0); );
  /// Input base voltage computation
w0:=eval((ut(i)*ln(i/isO(i))+eg(i))
....*nc+i*(re+rb/beta(i)), i=y0);
z0:=eval(tdev(i), i=y0);
Y0:=array([y0]);
W0:=array([w0]);
  /// Temperature computation in °C
Z0:=array([z0*T0-273]);
writedata(fy,Y0,float);
writedata(fw,W0,float);
writedata(fz,Z0,float);
Iin:=Iin*1.259;
end do;
fclose(fy);
fclose(fw);
fclose(fz);
```

Appendix B
Small-Signal Analysis of a Common-Source Power Stage

B.1 Introduction

The schematic in Fig. B.1 depicts the small signal linear schematic of a common-source power stage featuring a bias network that is modeled by the admittance $y_{g,b}$. In practice, $y_{g,b}$ may stand for a simple resistor or a more complex active load such as a *diode linearizer.*

The input and output currents and voltages are related to each other by an admittance matrix expression in the form:

$$\begin{pmatrix} i_g \\ 0 \end{pmatrix} = \{ Y_{PA} \} \cdot \begin{pmatrix} v_g \\ v_d \end{pmatrix} = \left(\{ Y_{unilateral} \} + \{ Y_{feedback} \} + \begin{pmatrix} y'_g & 0 \\ 0 & y'_l \end{pmatrix} \right) \cdot \begin{pmatrix} v_g \\ v_d \end{pmatrix} \tag{B.1}$$

where $\{ Y_{unilateral} \}$ is the unloaded unilateral admittance transistor matrix,

$$\{ Y_{unilateral} \} = \begin{pmatrix} 0 & 0 \\ g_m & 0 \end{pmatrix}$$

$\{ Y_{feedback} \}$ is the admittance matrix of the feedback network related to C_{gd}.

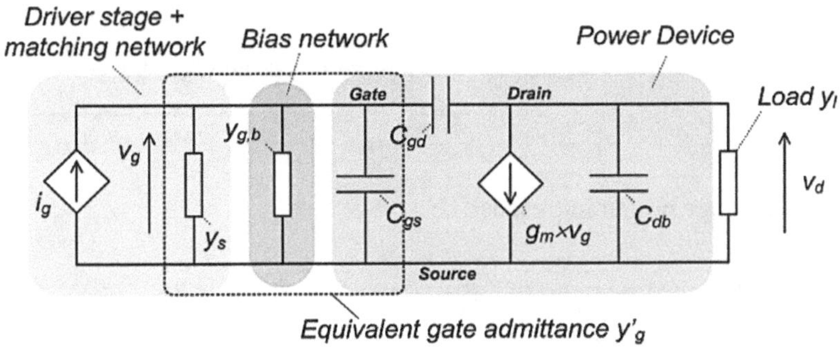

Fig. B.1 Basic quasi small-signal schematic of common-source stage including diode linearizer and non-linear sources

$$\left\{ Y_{feedback} \right\} = \begin{pmatrix} jC_{gd} \cdot \omega & -jC_{gd} \cdot \omega \\ -jC_{gd} \cdot \omega & jC_{gd} \cdot \omega \end{pmatrix}$$

$$y'_g = y_s + y_{g,b} + j\omega\, C_{gs}$$

$$y'_l = j\omega\, C_{db} + y_l$$

g_m is the linear small signal transconductance gain of the power transistor,
C_{db}, C_{gd}, C_{gs} are the drain/bulk capacitor, the gate/drain capacitor, the gate/source capacitance respectively,
$y_{g,b}$ is the equivalent admittance of the bias circuitry,
y_s, y_l are the input source admittance and the output load admittance respectively.

Therefore, the intrinsic admittance matrix of the stage becomes:

$$Y_{PA} = \begin{pmatrix} y_{11} & y_{12} \\ y_{21} & y_{22} \end{pmatrix} = \begin{pmatrix} j\omega\, C_{gd} + y_s + y_{g,b} + j\omega\, C_{gs} & -j\omega\, C_{gd} \\ g_m - j\omega\, C_{gd} & j\omega\, C_{gd} + j\omega\, C_{db} + y_l \end{pmatrix} \tag{B.2}$$

In the following development, a few assumptions will be made. Firstly, the transfer gain y_{21} is approximated to g_m. Secondly, in the band of interest, the output load y_{22} resonates and is approximated to $y_{22} \sim y_{l0}$, where y_{l0} is the real part of the output load admittance.

B.2 Power Stage Input/Output Admittances

The power stage input impedance in the band is:

$$Y_{in} = \frac{i_g}{v_g} = (y_{11} - y_s) - y_{12} \cdot \frac{y_{21}}{y_{22}} = y_{g,b} + j\omega\, C_{gs} + \underbrace{j\omega\, C_{gd} \cdot \left(1 + \frac{g_m}{y_{l0}}\right)}_{\textit{Miller effect}} \tag{B.3}$$

The power stage output impedance is:

$$Y_{out} = \frac{i_d}{v_d} = (y_{22} - y_l) - y_{12} \cdot \frac{y_{21}}{y_{11}} = j\omega\, C_{db} + \frac{g_m}{1 + \frac{C_{gs}}{C_{gd}} + \frac{y'_s}{j\omega\, C_{gd}}} \tag{B.4}$$

One useful condition of stability consists in sourcing the power stage with such an input admittance that the real part of the output admittance $\Re(Y_{out})$ is positive, which is equivalent to writing:

$$1+\frac{C_{gs}}{C_{gd}}+\frac{\mathrm{Im}\left(y_s'\right)}{\omega\, C_{gd}}>0$$

In practice, C_{gs} is a dominant term in the gate impedance, and the output admittance can be approximated as follows:

$$Y_{out}\approx j\omega\, C_{db}+\frac{g_m}{1+\dfrac{C_{gs}}{C_{gd}}}\approx j\omega\, C_{db}+\omega_T\cdot C_{gd} \tag{B.5}$$

where ω_T is the transition frequency.

B.3 Power Stage Non-unilateral Trans-Conductance Gain

The non-unilateral trans-conductance gain is given by:

$$\begin{aligned} G_m &= \frac{i_d}{v_g}=y_{l0}\cdot\frac{v_d}{i_g}\cdot\frac{i_g}{v_g} \\ &\approx \frac{g_m}{1+j\dfrac{\omega\, C_{gd}\, g_m}{y_{l0}\, y_g'}} \end{aligned} \tag{B.6}$$

B.4 Power Stage Trans-Impedance Gain

The composite trans-impedance gain is given by (assuming $y_{11}=j\omega C_{gd}$):

$$\begin{aligned} A_{ixr} &= \frac{v_d}{i_g}=\frac{-y_{21}}{(y_{11}\cdot y_{22}-y_{12}\cdot y_{21})} \\ &\approx -\frac{g_m}{j\omega\,\left(C_{gs}+C_{gd}\right)\cdot y_{l0}}\cdot\frac{1}{\left(1+\frac{g_m}{y_{l0}}\cdot\frac{C_{gd}}{C_{gs}+C_{gd}}\right)} \\ &\approx -\frac{1}{j\, y_{l0}}\cdot\frac{\omega_T}{\omega}\cdot\frac{1}{\left(1+\omega_T\cdot C_{gd}\cdot y_{l0}^{-1}\right)} \end{aligned} \tag{B.7}$$

B.5 Power Stage Transducer Gain

The transducer gain is:

$$\begin{aligned} G_T &= 10 \cdot Log\left(\underbrace{\frac{v_l^2}{2 \cdot \Re(y_{10}^{-1})}}_{delivered\ to\ load}\right) - 10 \cdot Log\left(\underbrace{\frac{i_g^2}{8 \cdot \Re(y_s)}}_{available\ from\ source}\right) \\ &\approx 10 \cdot Log\left(4 \cdot |A_{ixr0}|^2 \cdot y_{l0} \cdot \Re(y_s)\right) \\ &\approx 10 \cdot Log\left(4 \cdot \frac{\Re(y_s)}{y_{l0}} \cdot \left|\frac{\omega_T}{\omega} \cdot \frac{1}{1 + \omega_T \cdot C_{gd} \cdot y_{l0}^{-1}}\right|^2\right) \end{aligned} \tag{B.8}$$

Appendix C
Theory of Power and Volterra Series

C.1 Power Series

A common way to model non linearity in a memoryless power amplifier consists in using power series. In such a formalism, the output voltage is simply related to the input voltage by means of the following expression:

$$v_{OUT}(t) = \sum_{j=1}^{\propto} \left(a_j \times v_{IN}^{j}(t)\right) \tag{C.1}$$

where v_{OUT} is the output voltage, v_{IN} is the input voltage, and a_j is a real factor.

C.2 Volterra Series

Meanwhile, such expression does not take into account either memory effects nor phase shifts due to non-linearities (*AM/PM*). To this end, Volterra formalism is found to be an accurate theoretical tool. A jth order Volterra kernel $g_j(t_1, t_2, \ldots, t_j)$ is a multi-variable complex function that makes the output of an invariant system dependent of the state in which it used to be at various dates in the past. In this case, the system is modelled in time domain by a series of convolution products as shown in (C.2):

$$v_{out}(t) = \sum_{j=1} \frac{1}{j!} \int_{-\infty}^{+\infty} du_1 \cdots \int_{-\infty}^{+\infty} du_j \times g_j\left(u_1, \cdots, u_j\right) \times \prod_{r=1}^{j} v_{in}(t - u_r) \tag{C.2}$$

In an analogous manner, the system can be modelled in frequency domain by a series of integral components as shown in (C.3). In this case, multi-variable complex Volterra kernels $G_j(f_1, f_2, \ldots, f_j)$ are employed.

$$V_{OUT}(f) = \sum_{j=1} \left(\frac{1}{j!} \int_{-\infty}^{+\infty} df_1 \cdots \int_{-\infty}^{+\infty} df_j \cdot \begin{pmatrix} G_j(f_1, \cdots, f_j) \times V_{IN}(f_1) \times \\ V_{IN}(f_2) \cdots V_{IN}\left(f - \sum_{r=1}^{j} f_r\right) \end{pmatrix} \right) \tag{C.3}$$

where $G_j(f_1, f_2, \ldots, f_j)$, $V_{OUT}(f)$ and $V_{IN}(f)$ are respectively the Fourier transforms of $g_j(t_1, t_2, \ldots, tj)$, $v_{out}(t)$, and $v_{in}(t)$.

Appendix D
Analysis of Stability in Power Amplifiers

D.1 Theory of Unconditional Stability

Unconditional stability of a *RF* system is obtained when both input and output reflexion factors Γ_{IN} and Γ_{OUT} have a magnitude below 1 whatever the source and load impedances might be (D.1).

$$\begin{cases} \forall z_L \,(where\ \Re\,(z_L) > 0)\,,\ |\Gamma_{IN}| < 1 \\ \qquad\qquad and \\ \forall z_S \,(where\ \Re\,(z_S) > 0)\,,\ |\Gamma_{OUT}| < 1 \end{cases} \tag{D.1}$$

where z_S *and* z_L are the normalized source and load impedances.

The reflexion factors of a system are expressed in terms of its *S*-parameters as follows:

$$\begin{cases} \Gamma_{IN} = \dfrac{S_{11} - \Delta \cdot \rho_L}{(1 - S_{22} \cdot \rho_L)} \\ \Gamma_{OUT} = \dfrac{S_{22} - \Delta \cdot \rho_S}{(1 - S_{11} \cdot \rho_S)} \end{cases} \tag{D.2}$$

where

$$\begin{cases} \rho_L = \dfrac{z_L - 1}{z_L + 1} \\ \rho_S = \dfrac{z_S - 1}{z_S + 1} \end{cases} \text{and } \Delta = S_{11} \cdot S_{22} - S_{12} \cdot S_{21}$$

By substituting (D.2) in (D.1), unconditional stability is expressed by a new triple condition (D.3):

$$\begin{cases} K = \dfrac{1 + |\Delta|^2 - |S_{11}|^2 - |S_{22}|^2}{2 \cdot |S_{12} \cdot S_{21}|} > 1 \\ 1 - |S_{22}|^2 - |S_{12} \cdot S_{21}| > 0 \\ 1 - |S_{11}|^2 - |S_{12} \cdot S_{21}| > 0 \end{cases} \tag{D.3}$$

where *K* is the Rollett factor

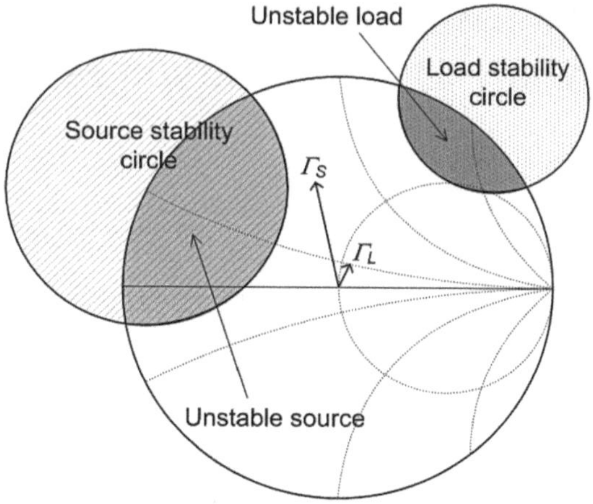

Fig. D.1 Smith chart illustration of source and load stability circles

Unconditional stability is hardly fulfilled in practice, especially for a common-source power stage that is little unilateral. In this case, stability circles need to be analyzed. The source (resp. load) stability circle represents the locus of the source (resp. load) impedances on the Smith chart that make $|\Gamma_{OUT}| > 1$ (resp. $|\Gamma_{IN}| > 1$). By the development of (D.1) and (D.2), it is found out that the radius of the source (resp. load) stability circle is given by r_S (resp. r_L) in (D.4) and its location in the Smith chart is determined by the vector Ω_S (resp. Ω_L) in (D.5)

$$\begin{cases} r_S = \dfrac{|S_{12}| \cdot |S_{21}|}{|S_{11}|^2 - |\Delta|^2} \\ r_L = \dfrac{|S_{12}| \cdot |S_{21}|}{|S_{22}|^2 - |\Delta|^2} \end{cases} \tag{D.4}$$

$$\begin{cases} \Omega_S = \dfrac{S_{11}^* - S_{22} \cdot \Delta^*}{|S_{11}|^2 - |\Delta|^2} \\ \Omega_L = \dfrac{S_{22}^* - S_{11} \cdot \Delta^*}{|S_{22}|^2 - |\Delta|^2} \end{cases} \tag{D.5}$$

The unstable source and load impedances in the operation bandwidth are illustrated by the darkened regions in Fig. D.1.

D.2 Practical Analysis of PA Stability

However, such mathematical formalism is found to be unpractical when a broad band *PA* design methodology is needed. The approach that is proposed in the following paragraph is based on a common-source power stage and copes with broad-band frequency operation and broad output *VSWR* variations. At its output node, this

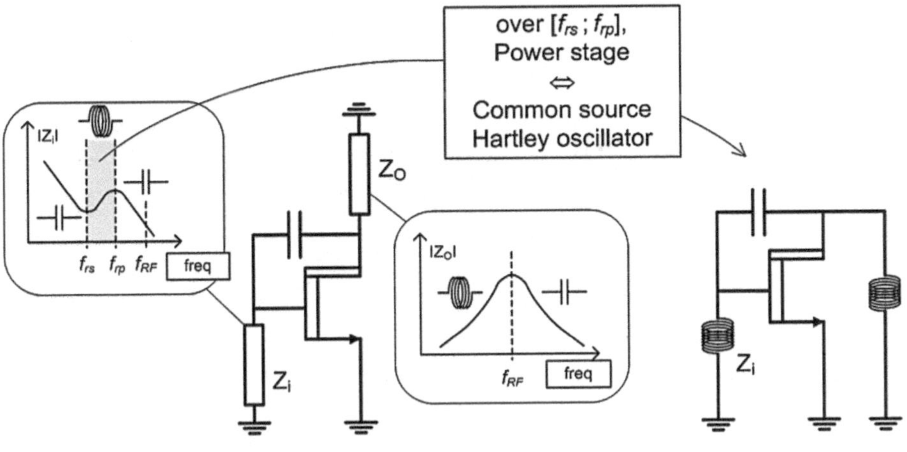

Fig. D.2 Illustration of instability for a common-source stage

stage is loaded by a pull-up inductor and a passive network (Z_O). This load overall resonates in the vicinity of the carrier *RF* frequency (f_{RF}).

Consequently, at "low frequency" below the carrier frequency, the output load is inductive. The input source impedance (Z_i) is more complex to analyze. At high frequency (above f_{RF}), it is essentially capacitive and is dominated by the power device gate/source capacitance. At very low frequency, it is mainly capacitive since it is determined by the DC-stop series capacitance that is necessarily integrated to connect the power stage to the driver stage. In between, the input impedance Z_i might become inductive according to the inter-stage matching network topology. Over these "inductive" frequency ranges (practically at medium frequencies), the power stage features the configuration of a Hartley oscillator as depicted in Fig. D.2 and is potentially unstable.

The *PA* should operate in a stable manner over broad output *VSWR* variations. This implies the load condition on Γ_{IN} in (D.1) must be fulfilled for any load impedance. In contrast, the source impedance is determined by the previous stages in the transmit path and is controlled with an acceptable accuracy whereby the source condition on Γ_{OUT} in (D.1) is loosened.

In order to improve stability (i.e. to reduce the Smith chart intersection area with the input source stability circle and/or the output load stability circle) without degrading drain efficiency, the simplest method consists in decreasing the input gate impedance quality factor by means either of a gate series resistor or of a gate shunt resistor (Fig. D.3). In the present book, the actuation buffers (see Chapter 2) or the *diode linearizer* transistors (see Chapter 3) stand for gate shunt resistors. The resulting input impedance Z'_i presents smooth resonances and is dominantly resistive at medium frequencies as depicted in Fig. D.3.

Assuming the load impedance is in an inductive region (either at low frequency or due to output impedance mismatch), the real part of the input admittance is negative and is given by the Miller effect in (B.3) according to (D.6). Providing it is kept

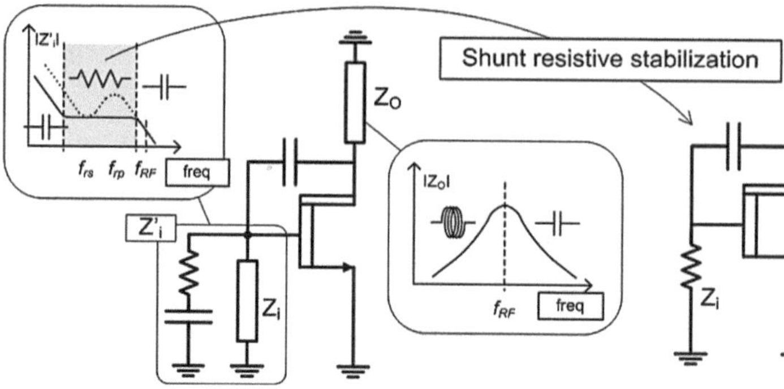

Fig. D.3 Illustration of stabilization by means of shunt resistive gate loading for a common-source stage

below the gate shunt admittance $\Re(Z'_i)^{-1}$ over the targeted frequency range, one ensures that $|\Gamma_{IN}|$ remains below 1.

$$\Re\,(Y_{in}) = -g_m\,\omega^2 L_O\,C_{gd} \;<\; \Re\left(\left(Z_i'\right)^{-1}\right) \tag{D.6}$$

where L_O is the equivalent output inductance.

Index

FSC
www.fsc.org
MIX
Papier | Fördert gute Waldnutzung
FSC® C083411

Kolibri 360 GmbH
Ferdinand-Jühlke-Straße 7
99095 Erfurt, Deutschland
produktsicherheit@kolibri360.de